The Welding Business Owner's Hand Book

How to Start, Establish and Grow a Welding or Manufacturing Business

David Zielinski

ISBN-13: 978 -1484045237
ISBN-10: 1484045238
Library   of Congress Control Number: 2013909134
CreateSpace Independent Publishing Platform, North Charleston, SC

CHAPTER 11 - BUSINESS PLANS AND WHY YOU NEED ONE 194

CHAPTER 12 - HIRING HELP AND EXPANDING YOUR BUSINESS 202

Introduction

As this book was written we made assumptions about the reader. As a starting point, we are presuming that the reader is a welding student or a journeyman welder interested in what managing a successful welding business requires, and not a business person who knows nothing about welding.

The goal of this book is to teach the welder how to think and act like a business owner. Perhaps the welder plans to invest in a business future beyond just going to work and earning a paycheck. If so, this is the how-to book, it is a blueprint covering every area of building a business in a competitive environment.

The reader will learn all the components needed to build a business, i.e.; marketing, business plan, management, networking, sales, competition, etc. Welders who study and learn the ideas, concepts and tools discussed in this book should be able to start-up a business. If, over time, those tools are mastered, the business may become quite profitable. Not only will welders benefit from this book, but experienced business owners will learn new ways to be successful. What this book offers is no different than learning how to weld. If you get the right information and practice long enough, you will have gained the skills required to earn a great income. Just like learning how to weld, you will notice this book has a lot of repetition. Practice makes perfect and key point of information are stressed so that the reader gets used to the ideas and concepts sooner.

This book is not for business magnates or investors looking to get into the welding, fabrication, machining or manufacturing industries! Welding is a unique industry that requires knowledge of specialized techniques that most other business owners cannot relate to. For example, what may seem to be a very simple job, such as welding a small pipe located in a particularly tight spot, may actually take an entire week to weld ***properly***. This is the reality of welding. On the other hand, a job requiring the installation of a free-standing, million pound structure can be completed in just a few hours by comparison!

Stay away from welding-related industries, if you do not understand the metallurgical issues that come with fabricating metal products! These industries require processes that cannot be rushed, or in other cases 'breezed through'.

Over time, the author and many of the welders and fitters he worked with made the transition from welder to welding business owner. The author studied successful business operations from some of the most successful and established welding shops and chose his business focus from among the large number of opportunities that welders have (see chapters 4 and 8). This book provides real world solutions to the problems that they faced when building a business. It is chock full of business do's and don'ts, and trust that there are lots of both!

The following are examples of comments from welding business owners who received advance copies; "This book is excellent! I wish I had this when I decided to go into business years ago." "It would have saved me a years of wasted effort." "I found that I had so much more to learn about marketing."

It is welders' personalities that draw them into this high-paying business. They heard of successful, high-flying welders somewhere in the world making thousands of dollars a week. Welders today, just as the gold seekers going to California during the nineteenth century, travel far, and work hard for the lure of big money. They travel the whole world for jobs becoming what they call "road warriors", risking health and well being, spending their own money, and doing just about anything for a fat paycheck. They work hard to get the really big money jobs. For example, government defense contract jobs.

They are risk-takers, they tend to be people who like to try new things, do not mind failing, and love to learn. They learn from their mistakes, and they don't beat themselves up when they make mistakes. As a personal example, the author of this book learned how to work on Wall Street, as well as a pipe welder. Both jobs were sought out for much the same reasons; the large paycheck. His evaluation of both sets of co-workers found them to be much the same people, different class, but the same attitude and mind-set.

The character traits of welders are; hard work, love of money and a never quit attitude. Those traits make them, in many cases, successful when starting a business. As business owners they are considered blue collar Ph.D.s for their mastery of many different business-building skills. They learn quickly that there are different types of education.

CHAPTER 1 - ARE YOU READY TO START A WELDING, FABRICATION OR MANUFACTURING BUSINESS?

If you are considering starting a welding business, this book will provide all the information you need. But, please consider these two most important facts.

1. Most welding businesses fail usually through some fault or oversight of the owner. *The reasons are explained later in this chapter.*

2. Once you become a welding BUSINESS OWNER you are no longer just a welder! First and foremost, you must now eat, breathe and sleep the life of a business owner. It is a big change, and unfortunately not all are able make this necessary transition.

The reality is that new business owners will have many new responsibilities and skills to master. They must do whatever it takes to get business in the door. That does **not** include anything to do with welding. In time, with practice and commitment, they can *become* competent in the many areas needed for success. They develop their talents, their business skills and a different mind-set. **Right here and right now, you need to understand that this piece of information is worth the price of this book, and could save you tens of thousands of dollars.** If you want to succeed in starting a welding business then you must be focused on a great many different tasks that are unrelated to welding, these include:

- Sales
- Marketing
- Networking

It is a fact that successful Welding, Fabrication, and Manufacturing business owners *ARE GOOD SALES PEOPLE*!

You might ask 'why'? You won't be welding anything unless you have closed a deal and have a paying customer who wants your services! Success depends upon the welding business owner becoming very competent in several very different job skills, responsibilities and keeping their eye on their ultimate goals. Welding business owners spend their time marketing, knocking on doors, cold calling strangers, and doing sales and administrative work.

Welding is no longer a priority. Yes, you will make money welding, but you need to get the work first. If you do not get business in the door, you will fail. That is what running a welding, or any business, requires and it seems as though that fact is never discussed. You need to ask yourself why that is? Most books and business services, claim they only want to be positive, and do not want to discuss the negative sides of business. The reason is simple. They wish to sell the new business owner unneeded services, and that is the truth! The real question is, "Are you willing to focus on sales, marketing and networking for at least the first few years?"

Your new business demands that you become an effective and aware business owner, while welding and fabricating at the same time. Those new obligations coupled with unfamiliar and stressful work responsibilities can make aspiring welding business people vulnerable to predators.

Once you are a new business owner there will be predators who want a piece of your business. They come in the form of "helpful services", and have very good reasons, as to why what they are selling is going to help you make money. They acquire your personal information from sources that list all the newly established businesses, and they prey on them specifically. All of their, "helpful services" are geared to take as much money from you as possible, before you go broke and/or give up. These "services" come in all forms and types.

You must focus on, keeping your expenses low, and meeting the right people. As far as these predators are concerned, the newer you are and the less you know, the easier their sale. They are professional sales people preying on new business owners who are looking for quick answers to their problems. The sales people do this by promising the customer the answer to any difficulty. Do not buy into the hype or urgency. Urgency is the main sales technique they use. They will say there are only two spots left and one is already gone. Get used to the high pressure sales pitches, or sob stories. They come across hundreds of people like you every day, and it is their job to close you.

Just remember this; every sales call has a winner and loser. They are either closing you or you close them. As a business owner you must accept the fact that you are in a financial war. This is how the business world works. The information you are going to need is in this book and you must put in the will to succeed and not give up.

Why Most Businesses Fail and The Reality Check You Need.

Allow me to illustrate with an example. We observe a welding business owner making tons of money, but not producing a good product, or good welds or anything else. Most of us may know a business owner like this and we don't understand why they are still able to do so well. Let's call this owner Bob.

Bob is always busy, pays his welders poorly, treats them badly and does bad work in general. The big secret is that he is a good salesman. He is cheap and knows how to take advantage of every situation. He is a successful business owner who can make it, no matter what business he has chosen. This is the exactly the kind of shrewd attitude and outlook that you need to strive for. Bob's business will always be successful because he knows how to get the work in the door, and he respects the value of money. You need to learn what Bob's good points are and use them to your advantage. These include knowledge of the nuts and bolts of managing his business, focus on his market segment, keeping his costs low and a laser-like focus on sales.

When most welding businesses fail, it is usually the owners fault. The leading reasons most fail come from three areas:

- Starting without enough cash
- Unwillingness to dedicate most of their time solely to Sales, Marketing and Networking, not welding
- Investing your time and money in ALL the wrong places

Starting Without Enough Cash

The first word that must be learned is *__plan__*. To successfully raise money for your business, you need a plan. You should also have a plan B in case you do not do it right the first time around. You must figure out what your business needs to reach your goals. Decide on your sources for money. Let's address the issue of having enough cash on hand.

If you don't have the cash then you are wasting your time and money on an unattainable dream. Most businesses fail because they cannot make it long enough for the owner to learn how to get business in the door. Starting out with the needed cash is the first part. You will need enough money to get yourself through the startup period of living costs and business expenses. (At minimum you need to include all of your living expenses, start-up monies for the business and enough cash on hand to survive at least six months without a pay check)

One of the biggest mistakes new business owners make is spending on new equipment. Stay away from new equipment because you can't afford it. You need to preserve your money for as long as possible, spending only for necessary expenses. It is most important that the business owner knows, where their money comes from, and where it goes. Remember, a 20 year old pipe bender bends just as well as a new one!

Unwillingness to Do Sales, Marketing and Networking

Many business owners open up their doors and expect people to just come in and spend money. If you open up a shop, and hope for work just because your doors are open, then you deserve to fail. If it were that easy, why would anybody work for anyone else? Unions and big companies would all be out of business because the little person would no longer need them.

The facts are that most business owners fail because they feel their work, products or services are so exceptionally good that they do not need to advertise or make personal sales calls. This is a death sentence in the business world.

To succeed you need to focus the majority of your

time, money and efforts on marketing, sales and networking. **A welding business with no customers is not a business. Customers are the focus of every successful business, so finding new ways to get their attention and build long-term relationships for your mutual benefit is of the highest importance.**

You must understand your customer's needs and seek to provide the best customer experience that you can. It will take a lot of work on your part, but it is the only path to having a successful business. You must know your customers and involve them in your business through educating them on why your services are valuable. The long-term relationship that you build with marketing must be built on trust.

Investing Your Time and Money Improperly

This is an area where many new business owners make a big mistake. Don't waste your time and money on things that do not work such as general advertising and spending your time on the wrong people. Many mobile business owners that I know complain that they have no work. However, all of them spend their days talking to other people who are in the same business. Don't spend your time and money associating with the people who are fighting for the same jobs that you are. Yes, you have a lot in common, but that is how you go broke and end up working for someone else. Idle chatter may be fine, when you are working for someone, but not when it is your business and you are in competition for work. Instead, get out there! Take a road trip, pass out some flyers, spend three days a week introducing yourself to business owners that may hire and pay your bills. That is how it is done!

Successful business owners make sure they can afford what they are doing by spending their time wisely while taking advantage of those who are not willing meet the people who ultimately will give you the work needed to succeed. The owners who fail buy shiny new equipment, pray for work and finally just complain the economy is just that bad for everyone. That is why 20% of the people in the world make 80% of the money. What side of that proven number are you on? Think about this before any business expenditure.

You have eliminated your boss! You are your OWN Boss now! What Are Your New Responsibilities?

So you are the new boss, and must learn your new responsibilities. You must learn to be a:

- Sales person
- Self-starter
- Productive manager of time
- Know the Law
- Welder, Fabricator or Engineer
- Jack-of-all-trades

<u>Every business owner is a sales person but with the responsibilities of a welder and fabricator</u>. They must be a self starter who knows how to manage their time. Time management is vital because, there is no longer a boss to tell them what to do. The new business owner might waste their days allowing their time to manage them. Don't get caught up in a false hope of work will find you. Just remember this, if you are not meeting people, getting contracts in the door, then you are not doing your job correctly. At that point your time is managing you. The successful owner practices self-control, and carefully manages both time and money until enough work is found. Only then can the costs of equipment, rent, supplies, leases, labor, and personal expenses be justified.

CHAPTER 2 – LEGAL ADVICE FOR THE BUSINESS OWNER.

Yes, you probably *will* need a lawyer for your business.

Remember, what you don't know, can and will hurt you. The law does not offer the same protection to a business owner that it does to a consumer. As a consumer you have many rights and businesses must ensure they do not take advantage of you. When you start a business all of your deals come down to what is in the contract. Whatever agreement you make with someone verbally, means little, unless that is the exact wording in the contract, and it is signed by both parties.

A common practice of established business owners is to get new businesses to agree to something and then sign another agreement in a contract. They know that you need the work and will, without delay, sign on the dotted line. They take advantage of you, and you legally have no right to anything, other than what you have signed for in the contract. It all comes down to what a judge would rule and that is always what is in the contract. Remember, when you rent a space for your business, commercial realty is like the Wild West. You do not have the same protection that you would if you were involved in residential real estate.

When it comes to running a business you will encounter every scam that exists. A lot of them come from some big names that you are familiar with, and perhaps already trust. The business world is a war where your enemies always come to you with a smile, and most importantly, a great story as to why you need to do what they say. If you can't handle that, then you need to forget about owning a business. This is the real world, and not the propaganda that we all learn about how respectable business people are. The only respect they have is for money.

Business is business and nothing more. Be friendly but always alert to tricks, scams and lies that will be used to get you to agree to an unprofitable contract, or a bad lease.

Business Structures

Since you are starting a business you will need to know what types of structure options may be formed. The types of business are:

- Sole Proprietorship
- Limited Liability Company (LLC)
- Corporation

Here is the deal. These are your choices for your business structures. What you need to do is find either a lawyer or a legal service to help you decide which one is best for you. Spend the money to get the legal help you need to make this very important decision! Trust me! You will avoid much costlier problems later down the road should you make the wrong choice for your business now.

Sole Proprietorship

Sole Proprietorship is when you work as an individual but also a business in terms of no longer being an employee. That means no benefits, no unemployment tax, no insurance nor the rights of an employee. The upside is you are just a person working as an independent contractor. The down side is if you mess up then you can be sued and may be liable for your mistakes for the rest of your life. Also, you are responsible for paying income tax.

Limited Liability Company

The Limited Liability Company, also known as an LLC, is the middle of the road. It offers the protection of a business but passes through the income to the individual. If you don't earn anything in a year then you don't need to pay a minimum tax.

Corporation

A Corporation is the ultimate form of protection. Your company would technically become another person. This means that whatever your corporation does, even though you are running it, you are NOT liable for the corporation's actions. The main thing to get if you open a corporation is an indemnification agreement that says you are in no way liable for anything that you have done. That is the law. The downside of a corporation is after the first year you may need to pay a minimum income tax even if you have not earned a penny.

The structure of the business you choose is up to you, your accountant and lawyer. Local lawyers do a great job but you can also file for the business structure yourself in your state. The other option is going with a legal service that knows the ins and outs of business and does it at a fraction of the cost. Here are some discount legal services that you can consult about your business structure.

www.AmeriLawyer.com

www.Incorporate.com

www.LegalZoom.com

www.MyCorporation.com

www.LawDepot.com

Financing, Credit Lines and Credit in General

- It takes money to make money

- Determine exactly what you need to reach your initial goals.

- Plan for several stages of growth, but budget carefully. Decide on sources for financing. Compare your progress against your projections.

If you are starting a new business financing will be very difficult. Banks, lenders and credit card companies only deal with well established businesses and need to see income and contracts in your hand before they will even talk. They want to see stability, cash flow and assets that they can sell real fast if things do not work out. The chances of financing a new business are not very good.

There is one way around this. Personal resources or collateral where you are responsible for the loan. If the business fails you give them your house, car or whatever you have that has any value. <u>The reality is that 80% of all businesses will fail in two years or less and with a number like that you would not be willing to lend your money either.</u> Small companies used to have access to private investors, but now those small companies are competing with large corporations seeking capital.

However, it is possible that your local welding supply store will most likely offer a small credit line. These credit lines are for buying equipment and or supplies. If you are a new business, then you will want to get a Dun and Bradstreet number to start building credit. Business credit is not like personal credit and you need to personally report the payments (no one will do it for you) you made for products and services. You can find out more by visiting Dun and Brad Street at **www.dandb.com**. As a side note, many of the legal services websites listed in the previous section offer well-established business structures for sale (you change the name to whatever you want later). Or they can set up a Dun and Bradstreet business credit identification number to start establishing credit.

If you have an existing welding business or are buying an established welding business then there are many options for borrowing money. Banks, lenders and creditors will be more than willing to talk because if you fail they can get their money back from the sale of the business.

Buying or Investing in Equipment vs. Renting Equipment!

As stated in the last chapter, <u>one of the reasons most businesses fail is because they do not have enough money to make it through the start-up period.</u> The biggest expense is equipment. A welding, fabrication, machine or manufacturing shop needs a great deal of money to set it up properly. Do not purchase equipment unless you are using it daily, or have a contract for a job that requires it.

Be aware that those who sell new and used equipment are also predatory in their business practices. They have experience selling equipment to those who have a dream of owning a business, while knowing that within a year or two they will be buying it right back from them when the dream fails. Business is a survival game made for those who have the most self control. What are you to do in this situation if you need equipment? Here are three options besides buying new or financing equipment:

- Rent
- Partner
- Buy used

Renting Equipment

Renting equipment is expensive; but the cost is a lot less in the short term than buying it. If you only have one job, a single rental payment makes a great deal of sense, and saves a lot of money. Renting works great, and keeps your cash free for other things. There is another option for you to get access to equipment.

Many welding shop owners are struggling, unable to make their rent or mortgage payments; you can rent their equipment, shop, or even their personnel. It is done all of the time and the owners who are renting use the extra money you pay them to remain in business. You can rent an entire established shop with all of the bells and whistles that you will need for just a few hundred or thousand dollars for an entire month, or for a single job. All you need to do is ask around. A lot of shop owners will be very glad to hear from you!

Partnering Deals

Partnering deals is another option. If you get a job and don't have a shop, equipment, or personnel, then partner a deal. You can take on bigger jobs and everyone is happy. Most welders are not good sales people so partnering with someone who is, is usually welcome for most people. It is in many cases a financial life line to the struggling welding business owner. Just be sure to get your partnership deal down in writing. Once again, it would be smart to have a paralegal or attorney take a look at your agreement before committing to anything.

Buying Used Equipment

Finally, if you need to buy equipment, and money is tight, and you do not have guaranteed work to pay for it, you may purchase used machinery. There are always welding shops shutting down because they can't get the business in the door or spent all of their money on shiny new equipment. Now you can buy that equipment for pennies on the dollar. You just need to know where to look and we're about to tell you!

A $5,000 pipe bender that is less than a year old might sell for $1,500, and accessories that cost thousands, might sell for less than $100. These types of deals can be found daily, but you have to look for them. However, buyer beware! People selling this equipment are opportunists living on the former welding business owners that didn't make it. They can smell desperation from miles away and will do everything they can to get every penny out of you. Sometimes though, you might actually find free equipment on these websites:

www.Craigslist.Org

www.IndustrialMachinery.com

www.MachineTools.com

Or just type in "Used Metal Working Equipment" into a search engine and you will find more than you will know what to do with.

Don't forget, whatever you choose to do (buy used, rent, lease, partner or any other deal you may make) sign a contract, detailing the exact terms that everyone agreed to, and then have your 'legal team' take a peek at it first. This is a critical step that should never ever be overlooked or cut to spare expenses.

Dealing with Legal Contracts, Insurance, Taxes and Leases

<u>If in doubt, call your lawyer out! That is a great piece of advice!</u> Don't pretend to know what is up or go along with what everyone else is saying. Business is a financial war disguised as a friendly encounter. Always have your lawyer handy.

As for taxes, you must have an accountant. Reading books for this information is a waste of time and money. Laws and rules change quite frequently, and rapidly, so you need the right people to protect you. You don't have the time to spend on these things. Hire professionals. That's what they are there for!

Your lawyer and accountant can advise you on all of these issues and in the worst case, refer you to the right person. In the business world legal mistakes can cost you your business and future income.

Before you even get any work you will need to find a contract and the proper legal documentation to sign. Once you buy your contract software or book then this is a good chance to meet some local lawyers. Lawyers will be willing to give a free consultation as long as you tell them that you are a new business owner and are in need of having a lawyer on hand for future issues. Another option is if you use a legal service like LegalZoom.com or AmeriLawyer.com they offer a yearly paid legal consultation services. You pay a yearly fee of about $100 and you have access to legal advice for the year. This service is good for what you are paying, but it is nowhere as good as having a local lawyer on call. This is also a good time to meet an accountant and find out what you will need and how to set-up your document filing system. In the end just remember ... <u>When in doubt, call your lawyer out</u>!

How to Deal with Partners, Friends and Family in a Welding Business

This is a touchy subject and it needs to be addressed correctly. Since this is a business everything needs to be on paper. Best friends, family and lovers all want to help but there is a time where you need to separate ties and emotions for the greater good. Everything is fine until you become successful due to you reading this book. In reality, your success is up to you and the choices you make. Sooner or later, friends and family will inevitably become jealous when you make it big. Be upfront and honest throughout the entire process. Clearly state in writing, all of the things you want and all of the things they want out of the deal or partnership. Get a contract signed and make it crystal clear as to what is expected. If they pull the *"but we are family or friends"* story, then tell them you don't want to risk the relationship and that is why we should all agree upon a deal on paper **before** anything happens. This is how marriages and most partnerships go bad, and why low income people never get ahead.

Rich people know what is expected because everything is in a contract. Besides that, they stick together. Poor people all want to help each other until one person gets ahead, then, in many cases, jealousy creates a situation that will undermine and destroy the successful business owner. Best friends suddenly become the worst of enemies. If you want to avoid the situation, with legal help, write a contract and don't assume anything.

CHAPTER 3 - ANALYZING YOUR MARKET, STRATEGIC PLANNING, COMPETITION AND CHANCES OF SUCCESS

Knowing Your Market, Yourself and Your Competition

Every day in a welding business will offer something new to learn or to do. The welding business owner must answer these questions:

- What do you have to offer?

- Do you have the talent to make it better?

- Can you create new and useful products?

- Should you create a niche business or have a general welding shop?

- Should you buy an already established business?

You must know your market, your competition, and yourself to be successful.

Knowing your market, and everything about it, is the key to any business. You need to determine what you will be competing for and who your competitors are. Every business environment is different, thus there is no one-size-fits-most welding business plan that works everywhere. So, where do you start? Here is a suggestion or two:

- www.Wikipedia.org
- Look up your state, county and city or towns websites.

Study your local economy and determine what is unique about your area. This is a thinking process that requires analytical skills and an open mind. Get to know your potential customers, the market you will be in, and your competitors. Let the market tell you what to do and do not force your opinion where the business is. Searching for water in the desert is a waste of time. Smart business owners listen to potential customers and study their market. Listen and look for the signs as to where you can find work for your business. Sit down and determine your strengths, your weaknesses, business opportunities and what is out there that may hurt you.

For example, I live in South Florida and the economy here is not good for welders because we have no real industry or manufacturing base. What we do have a lot of is rich people. Rich people have yachts, huge houses, personal jets, etc. and they love to spend money.

There is no industry or welding jobs here except in the area of tourism. You can try to run a pipe line rig welding business or fabrication shop but you will fail. If you want to survive here then you need to think about a service industry that targets the wealthy customer. This is an example of knowing your geographic area and your market. These would be my choices as a person living in the ritzy South Florida area specifically:

- Mega yacht repair and modification
- Fabricating specialized, high-end marine equipment
- Architectural iron and art metals
- Remanufacturing and repairing aircraft parts
- Cruise ships service and repair
- Restaurant equipment repair and fabrication

If you live in an area such as rural Wisconsin then you would want to service the farming industry, and food processing industry. Some examples of the types of businesses that would work are:

- Farm equipment repair
- General weld shop
- Stainless steel dairy equipment fabrications and repair
- Silo tank fabrication, modification and repairs

These are just a few examples of opportunities specific to your geographic area. Study your potential market by going online, driving around and learn the industries in your town, county or state. It takes time and effort to figure out what would work for you specifically. Know your market, yourself and your skills.

Most welders believe that they know themselves and their talents, but in the business world it is not about welding skills or how well you fabricate. The question that is asked is, "Can you deliver the product or service on time and on budget?" For example, a business owner states he needs to hire a welding contractor to erect a 10,000 square foot store. This store requires a steel structure completed in two weeks time or less. **Before** you present the client with a contract proposal, you'd better be certain that you can complete the job within the required time frame and budget. Don't promise, unless you can deliver. Should you fail, you will burn your future contacts. Business communities are very small, and it's more than likely they know each other, so you need to know yourself and what you can deliver.

Finally it all comes down to location; what is available and what you can provide. You need to find an area and a local economy that works for you. It is a long term process that takes lots of trial and error, all while listening to and understanding your potential customers needs. Take your time and carefully analyze the local market. We are all different so there is not a single solution that will work for all of us. For example, some people just do restaurant equipment repairs and fabrication. For some reason restaurant owners just seem to like a welder doing the work. These welding shop owners just fell into their industry, not expecting that to be their business focus.

It is a combination of the services you provide, and your marketing and networking skills, that make business owners feel comfortable hiring you. It's not an exact science, but a trial and error process that is different for every business owner. Something as odd as a type of hat you wear, can appeal to a certain crowd. You just never know what is going to work for you. It is the equivalent of growing up and figuring out who you are in the business world.

What Do You Have To Offer?

Are you so good at welding that no one else can do what you do? The answer is probably 'no'. There are exceptions, but they are very rare, and the chances are they are not you or I. What do you have to offer that is new or unique? These are questions you need to answer as a welding business owner.

In most cases you will offer the same services as the next business owner. You need to stand out and this comes down to being a better sales person. Can you do the job better, cheaper, faster or both? For most of the welding jobs it comes down to cost and time. Nobody really cares about anything else. This is why your competition, who may be a great deal *less* skilled and prepared than you are, gets more business in the door than they can handle.

Your competitors price and time table promise were just right. This is an example of welding business competition that the new owner will face. The business owner must find his own way to win the business. One thing that can be done is to produce a very complimentary work history, and provide references that could be checked, even references from friends, photographs of work would be helpful. You must appear as a well organized, competent, highly professional welder.

Can You Make It Better?

What can you make better than anyone else? If you are a welder, can you make the best welds or a better product? Here is an example of an area that you can make it better. Chrome heavy wall pipe welds are big money! Really big money! We are talking about a 20" pipe with a wall thickness of 2" plus inches gets about a $50,000.00 price tag (not for you) from the general contractor or engineering company (the business that got the work, you know the 20%) for a good x-ray quality weld that takes a week to do. Most businesses struggle with them and hire outside contractors for a set hourly rate to make the welds. It is a high risk weld in terms of the welder's health and the weld x-ray failure rate on the open root. If you can come in and offer your services to major power plant builder for the right price then you would likely have work all around the world, simple servicing this type of welding. It's a tough gig but there are people out there that make a fabulous living doing this kind of work.

Can you improve a product? If so, then you might just have a solid idea and a stable work stream coming in for your product. Every product out there has already been made, but the new ones that make it are successful because they are either:

- Priced Lower
- More Useful
- Look Better

You need to think of yourself as the buyer of a product and find things that would makes someone's life easier and work on *that* concept to improve upon or create a new product! It is not easy but if you can do that it will pay off in the long run!

This is a creativity game that requires the most odd ball welders, creative fabricators and artistic people out there. Welding is an **ART form** but the majority of welders are not artists. Talented welders are like tattoo artists and surgeons, in that, they are both in high demand, but never available. Both are common but few make a real impression. The bottom line is to have a product or concept you can make that is unique and indispensible to your target audience. If you are still working on that, then you need to focus on either your sales efforts, marketing or underbidding your competitors to get the job.

Niche Welding Business or General Welding Shop

Niche means a very tight area of a market. It is like a general weld shop as opposed to a specialty shop that only welds Lamborghini Exhaust Manifolds and that is their niche. It is a specialty area that you have experience or skill in that few other people have. Now that that is out of the way here is the deal with niche business versus a general weld shop.

A general weld shop does every type of welding but a niche welding shop only does one type of welding or concentrates on a single product. Here are the pros and cons of both:

General Welding Shop/Business

A general weld shop has a general customer base. Not an exact science but anyone who needs a weld or wants to fabricate something will be coming to you. You get all sorts of work coming in the door and your success depends on just about every type of weld or product needed! You would advertise to many industries and have a wide range of skills and equipment for getting the various types of jobs done.

Niche Welding Shop/Business

A niche welding business would seek customers that have been looking for you, but who don't have many choices or competition for the specific welding service they require. This is an area in which very few people fail, but unfortunately most never even attempt focusing on a niche. A niche business does not get many customers or business in general, but when the client does show up the show owner can charge whatever they want because there is no competition. Case in point: a business has a $100,000 die that needs a crack repaired, but if the wrong welding business is chosen, they might ruin or destroy it. A general weld shop will probably do a good job fixing it, but the customer will not trust just anyone with such a large investment. They will go way out of their way to find someone who fixes broken dies, and nothing more. It's like when you go to your doctor and they say you need to see a dermatologist (just an example). A dermatologist has one niche and that is skin issues. Anything that has to do with our skin a dermatologist can fix. Are you the equivalent of a dermatologist in the welding field?

Another great thing about niche businesses is that customers will find you from all around the world if you know how to make yourself available to them. People naturally have more confidence in someone that specializes in the one type of service they need done.

Let's say you repair tools and dies. Chances are that people will ship their broken components to you no matter how far away you are because they feel safe in paying you to fix it. The last thing anyone wants to hear is "we have never fixed one of these before but we will give it a shot". Think about this, you go into the hospital and find out you need heart surgery and the surgeon says "I have never done this before but I will give it a shot". You would **run**, not walk, away from that hospital in a hot minute! Any halfway intelligent person would!

As for specializing in multiple niches you can always open multiple business names for those specific niches. Large corporations do this all of the time. They have specific companies that only bid for certain types of jobs and you can do the same thing! Why reinvent the wheel? It's working for the big guys, it most certainly can work for you as well!

Since you are starting a business then you need to think about these options, how they affect your business and which are a better choice for your personality! It really comes down to, are you good with new projects every day or do you do better doing the same types of welds every day? Just give it some time and things will work out for the best.

Buying a Established Business

Here is what you need to know about starting a business vs. buying an established one. Starting a new business costs 80% less because 80% of the businesses value comes from established customers that provide income for the business. This is the same reason you can't get financing and loans for a new business.

Banks manage their risk by knowing what their risks

are, however it must be said that banks have no ***real world*** idea about how to establish a market value for a business. They have a set group of requirements they check for (assets and current contracts in hand) in order to establish a value for your business, but they do not take into consideration things like your company's growth rate, product rights, trademarks, etc. Established businesses rarely fail, while new ones are almost guaranteed to go out of business within two years or less!

The same numbers of 80/20 are true for buying a business. 80% of established businesses will last and 20% will fail. It does not matter how you examine these numbers because they always end the same. One thing you will learn when running a business is that the numbers never change. There are hot streaks and cold streaks, but no matter what, the long term numbers are always the same. If you buy an established business then you may expect an 80% chance of success. If you start from scratch then expect a 20% chance of success.

It is a fact that an established business has products and/or services as well as repeat customers. As a start up, you need to draw in new customers which that takes time and money. If you choose to start a business from scratch you must ask yourself, "Do I have the time, money and commitment to being the kind of sales person necessary to make the business successful"? If not, can I afford to hire the right kind of salesperson?" Might the answer be for you to save up a good chunk of cash while working for someone and then buy an established business? It's definitely an important question you need to consider before investing your money in totally the wrong place. Is your goal to own a successful, money-making business? Or are you more interested in starting something from scratch even if it means you're more likely to fail?

CHAPTER 4 - DECIDING ON THE TYPE OF BUSINESS THAT YOU WISH TO BUILD AND UNDERSTANDING YOUR FUTURE MARKET.

You must turn your planning into behavior. Primarily, you need to know who or what comprises the particular market you want to target, and how exactly will you service that market?

For the customer, there are many types of welding businesses to choose from. For the welder, it all comes down to learning your area, knowing your local economy and offering the right services to people who need or want them. Here is a rundown of welding business types and basic ideas/options:

Starting a Independent Contracting/Sole Proprietorship Welding Service

Independent contracting is the easiest way to start a welding business. It does not require a lot of legal paperwork or monetary investment. You could start for less than $1,000, and be working by the end of the week. The typical job comes from a company who is already providing welding services or is already engaged in some type of manufacturing business. These are businesses that need help but do not want to hire an employee because of the insurance cost, salary and benefits they'd have to provide. The easiest way to describe this type of a welding business would be to say that it is more of a self staffing service than a welding business. Work is easier to get than a typical welding business and the pay is also a bit less, but so are the start up costs. The steps to becoming an independent contractor or sole proprietorship are:

- Get an independent contractor agreement signed
- Buy insurance
- Promote yourself (advertise your goods/services)
- Provide high quality goods and/or services
- Collect your fees quickly and pay your taxes in a timely manner

That is all there is to becoming an independent contractor or sole proprietorship welding service. An independent contractor is like being an employee except that you charge a higher rate and receive no benefits. As an independent contractor you will need to make sure you have an agreement signed by you and the company *(BEFORE you begin the job!)* that states the pay rate, the type of work to be done and when you get paid. This is extremely important because you will inevitably get stiffed on your pay if it is not included in the contract. Case in point, a company paid for my travel from Florida to California, in order to weld on a cruise ship in the Pacific Ocean. I had worked for this well-known company many times, so I trusted them. I did not have the agreement signed before getting on the ship. This was a big mistake.

We all signed our contracts after we were out at sea and never got our signed contracts back. This was a smart move on the company's part because now we were all at the mercy of the owner's generosity or lack of (a good guy but also a very savvy and very successful business owner), and he owed us nothing if he felt that he did not want to pay us.

I logged in 256 hours in 16 days (20 plus hour shifts with no days off) and lost my travel pay, and received a lower hourly rate than the amount I was promised. This was my fault because I should have had the agreement signed by me and the owner of the company **BEFORE** leaving my home. In the end, this mistake cost me about $2,500 in pay. I received about $6,000 for two weeks of work, but would have had much more if I had followed my own rules. I have no one to blame but myself.

As an independent contractor you do not have the benefits of a full time employee such as health insurance, vacation time, etc. There is contractor insurance and you need your own health insurance. You are personally responsible if you get hurt or destroy something while on the job. Get this straight! If you lose your hand, there is no disability insurance, unless you purchased it. You are a business now operating under your social security number and don't have the protection of the government, state or Department of Labor. There is no overtime pay and everything is a flat hourly, or per job rate. If you destroy or break something valuable, the company will go after anything you have that is of any value at all. In return for taking on the risk, you can charge whatever you feel is fair, as long as everyone agrees to the rate.

Once you finish the job you will need to collect your fees. This was specified in the signed contract. Some companies will pay weekly, others will make you wait between 30 to 90 days for your check. Once you get your money, you are responsible for paying your own taxes. You get a check for the full amount with nothing being withheld for taxes. Just a quick note, the IRS will know what you owe because at the end of the year you will be issued a 1099 just like an employee gets a W-2 form. Make sure you have enough money put away to pay your taxes or you will quickly find yourself in a lot of financial trouble you do not need.

Being an independent contractor is a good alternative and a half way point between being a full time employee, and owning a business. It is an excellent start for someone who wants to get a taste of being a welding business owner without the big commitments or overhead costs. This is also how companies hire welders and craft personnel in Europe and it is the standard in the cruise ship industry. It allows you to earn more money with just the bare minimum of investing in health and contractor's insurance, and a cheap set of business cards. In most cases you do not need any more equipment than you would need as an employee who is a welder. The basic safety equipment and clothes that you already have will be more than enough in most cases. Another note, since the company hiring you is no longer responsible for you, they are no longer concerned about your health and safety. It is now up to you to make sure you are safe. Trust me when I say that there *will* be job situations that you'll be faced with that are not only unsafe, but actually detrimental to your health and well-being. Be prepared by knowing how much risk you are willing to take before you decide to take a job. Welcome to the reality of the everyday business world.

Starting a Mobile Welding Services Business

Starting a mobile welding business requires a lot more commitment, money and equipment than an independent contracting service. Here is what is needed to start a mobile welding business:

- A business structure such as Independent Contractor, Limited Liability Company or Corporation.
- Your own welding equipment and tools
- A way to transport your equipment and tools to and from the job site
- Insurance
- Business cards and marketing materials
- Contracts
- Money to survive the start-up period

Chances are you will be working directly for the client if you are a mobile welding business. This means getting work is going to be tougher than being an independent contractor, but the up side is that you can charge more per hour. You are going to need some form of a business structure and a Federal Tax ID number. When dealing directly with customers they need this information for tax write offs and to satisfy the legal requirements of their legal department.

A mobile welding business relies on you owning a welding machine. What type you'll needed depends on what type of work you will be doing. In most cases you will need an engine driven welder that can do Stick, TIG and MIG welding plus some sort of cutting equipment. Once again, this all depends on the type of work that you will be doing.

For example, all you will need is a simple plug-in 150 AMP portable Stick welder with a small TIG rig attached if you are repairing restaurant equipment. If you are doing industrial repairs then you will need a pretty powerful engine-driven power supply that is enough for welding heavy plate and carbon arc gouging.

Transportation is the main attraction of a mobile welding business, and that needs to be chosen carefully depending on the type of welding services that you will provide and **where** you are most likely to provide them. You will want a truck that has four wheel drive if you are welding pipe lines or doing industrial construction. You don't want to be that contractor driving up to the job site that keeps getting his vehicle stuck in the mud or sand.

Let's say you are doing restaurant equipment repair or welding pipe on a ship then your car is more than enough with a 10 pound plug-in portable Stick welding power supply and some rods. The vehicle you choose needs to be suited to the type of work that you will be doing.

Finally, a strong warning **not** to use a mobile welding business as an excuse to lease or finance a shiny new welding rig or truck! You can get the right vehicle USED, or as they say now "preowned", for a fraction of the cost from the other welders who have failed. Be wise and take advantage of those welders who did not make it in the business by buying their equipment for pennies on the dollar.

Insurance for a welding business is pretty much the same as a independent contractor. You need contractors insurance to protect yourself in the event of something going wrong or someone suing you. You also need your own health insurance just in case you get hurt on the job, or get sick.

Any business that does not have customers coming in the door needs to have some type of marketing material. Business cards, flyers, magnetic signs for your vehicle, and anything that you can think of to promote your business to your potential customers. Don't kid yourself into thinking you will get jobs simply for being in business and having a sign on your door. You need to go out there and meet as many people as possible. That is the heart and soul of any mobile business. People buy from people they know, and that is how relations start that lead to signed contracts.

Create a list of all potential customers that you can contact within a sensible driving distance. Get on the road and visit them *all*, and leave them with a packet of your welding information. This is known as cold calling. As stated, get your foot in the door and sell yourself. Your motto should be *"quality service as the customer defines it"*. A hard and fast rule to live by... Produce better work than the customer expected at a reasonable cost, and you will be successful.

Once you get someone interested in hiring you then you need to negotiate a price, hourly rate and terms of the job. All of this is done in a contract that you can either buy in a book form, a software package, or a attorney can make one for you. Either way you should *never* do any work without a contract signed. Have I said that before? Yes! Because it is critical as a business owner that you never make this foolish mistake!

Ultimately, do you have enough money to pay for all of the business related items you just bought? Do you have enough to money put away to survive without a pay check for at least 6 months? If not, then you better have a part time job. Chances are it will take you a few months to start getting enough business in the door to pay your bills and collect your checks. During the first few months all you will be doing is meeting people and marketing. Once you get the work it typically takes between 30 to 90 days to receive your check from the invoice that you submit to the company at the end of a job. During this time even if you are working there will be no income.

Starting a Welding or Fabrication Shop

Starting a welding or fabrication shop is a large financial risk and a long term commitment. For most welders it would be a good idea to either start out as an independent contractor or a mobile welding business. Opening up a shop is not a huge risk anymore once you have some work and established clients. In order to start a welding shop you will need to have the following:

- A business structure such as Independent Contractor, Limited Liability Company or Corporation.
- Equipment and all of the tools needed
- Shop space (negotiate a lease)
- Insurance
- Business cards and marketing material
- Contracts and office equipment
- Money to survive the start-up period

Business structure depends on the type of work that you will be doing. At the very basic level, if you are just doing repair work and small scale welding, then working as a independent contractor/sole proprietorship is perfectly acceptable. On the other hand, if you are fabricating and providing services to larger businesses such as engineering firms, general contractors or biding on jobs, then you will need a Federal Tax ID number and either a Limited Liability Company or a Corporation. These are a must for certain types of clients and they will not even talk to you unless you already have this all set-up.

In most cases a shop needs a lot of equipment because you will be doing many types of jobs. This is another area that requires a large investment and commitment. Your three best bets are to either rent when needed, buy used, or the best option of using/leasing someone else's shop space and equipment on a 'per job' basis.

Shop space is going to be needed and this means that you will need to negotiate and sign a lease. The two options are leasing a space yourself or rent a part of someone else's shop until you get you foot in the door work wise. Just remember, most shop owners are struggling and would be more than happy to sub lease their shop floor to you. In most cases, if you get enough work they will be happy to partner the jobs, and let you even use their employees. If you sign a lease then you will be on the hook for the rent until the lease is finished. Use a lawyer to negotiate the contract and terms if you are signing a lease. Commercial leases are tricky and even a lot of Realtors don't like messing with them for good reasons.

Insurance for a shop is pretty much the same as a mobile welding business. Just ask your insurance agent and they will tell you what you need. In addition to insurance you will also need to comply with local and state laws. That means getting an occupation permit, fire inspection and anything else that your area requires. You need to visit your town or city hall to find out what is needed. You should also discuss these things with your lawyer to ensure that you aren't missing anything important.

As a business owner you need to get the work in the door. This means lots of marketing, networking and building relationships to get those jobs. Just because you have a shop and a big sign does not mean you will get enough work to pay the rent. This is a mistake that most shop owners make. They believe that simply because they are in business, people will bring in work for them to do. Some small jobs will come, but most of them are likely to be more of a pain than they are worth. It is your job as the shop owner to go out and bring back the types of contracts that will pay for the shop *and* put money in your pocket.

You will need basic office equipment to run your business if you are in a shop. It is also an investment that you need to make in order to get your shop established. Some of a shop's basic needs include a phone system, a copier/fax/scanner/printer, a filing system, etc. The positive side to having your own office equipment is that you can print and design your own promotional materials when you are starting out.

Finally you need enough startup capital to pay for the equipment, lease, and all other expenses and to make it long enough until you get a steady income flowing in. A shop requires a lot of commitment and savings to make it work. That is why in the beginning you should either start out as an independent contractor, or a mobile welding business and save up. Once you learn how to get the work and have established clients, then it would be smart to sub lease someone else's shop until you know for a fact that the investment is worth the risk.

Starting a Manufacturing Business

Starting a manufacturing business is a little less risky than starting a welding shop. That remains true as long as you already have a product that you can sell. A manufacturing business is just like a weld shop in every way except for needing to get work in the door. All of the requirements are the same except you may need product insurance and manufacturing approvals certifications such as UL (Underwriters Laboratories) approval. You can get all of the free help and advice you need for getting your products approved from the Small Business Administration at **www.sba.gov**. In many cases you may even be able to make your product at home or just rent some shop space for a week out of the month. The main issue with opening up a manufacturing business is selling the product, but for this type of business there are many options that will be covered later on in this book.

What Type of Business Should You Open?

The type of business you choose to open should be based on research and knowing the local economy and your customer base. What works in one area will not work in another. Sometimes choosing a type of business comes down to providing something no one else is doing which means no competition. Take your time, do your research, and make your decision based on what the customer tells you. Know your customer; involve them in your business.

Make a list of the positives and negatives for each type of business in your area and then make a logical, educated and informed decision on how you will invest your money. Keep in mind that a unique market segment, lower costs, better service is, in many cases, not enough. Do not try to one up the competition, but instead strive to achieve operational excellence. Make your business the best it can be, and keep in line with your strategies. Do your planning from the bottom up, not top down.

CHAPTER 5 - MARKETING FOCUS FOR YOUR BUSINESS

The Importance of a Business Image and Advertising

Marketing is the process through which you make your customers aware of your business and its services. Basically, you want to inform your potential clients of all the benefits of using your company such as how you do your business, and what happens after they buy your services. It is the part of your business that touches the customer.

Corporate Image

McDonalds, Sunoco, Exxon, NASCAR, Harley Davidson, Coca Cola, Pepsi and every other product and service you use and buy has an image, color pattern and logo that you recognize with just a quick glance. Think of a soda and chances are it is either Coca-Cola or Pepsi. If someone says Miller Welders you think the Power of Blue, Lincoln Welders you think red, and ESAB and HeliArc brands you think yellow.

These are successful business images and advertising campaigns. This is called "branding". You want to make sure your "brand" of product and/or services stick out in the minds of your clients as the one to rely on for **all** their welding/fabrication needs.

Creating a business image takes commitment, money and time. It is also a very powerful tool to get new customers to trust you with their projects. As a welder don't bother trying to do this yourself unless you are very artistic. In the past it did cost a lot of money to have a logo and promotional material made. However, today with companies outsourcing to foreign countries you can have an entire business image created for less than $100. If you were to go through an American company the same work and designs would cost well into the thousands. I am not a fan of outsourcing, but when you can get a $2,000 logo designed for $49, and it is just as good as the more expensive American-made image; that is a no brainer!

I had a few made for this price and the work is outstanding. As the perfect example, the cover of this book was done by **www.LogoNerds.com** for $29.95. That literally saved me thousands of dollars and I think this book's cover looks pretty professional and eye-catching. Don't you agree? You can get a whole business identity package for just $99 and then you just take the designs to your local print shop and have them print it. This is the new economy and as a start-up, it is a life saver.

When creating your business image you want to start with a logo and match your signs, letterhead, business cards, envelopes, promotional materials and uniforms to the same designs and colors. Everything needs to match and after some time you want people to recognize your business just by the colors or some type of unique design.

The bottom line is people recognize a business image and the sooner you start the sooner people will associate your business as the place to get welding work done. The main benefit of a highly recognizable business image is you can charge a higher rate than a business that is not recognizable because you have a trust established before ever getting the job.

Think of it this way, if you see a Harley Davidson logo on a product chances are you assume it is a quality item. That is the power of a business image and what you want your future customers to feel about your business. Here are some cheap places where you can have a business image made at a reasonable price:

www.LogoNerds.com

www.PrintforLess.com

www.PrintPlace.com

www.PrintRunner.com

www.UPrinting.com

www.VistaPrint.com

Reaching Your Customers

First thing you should think about is to focus on your

existing customers. Get more money to migrate to the bottom line by finding better ways to serve your customer base. Keep in mind that 20 percent of your customers bring in 80 percent of your revenue.

Whether you advertise or not the bulk of your marketing efforts and dollars probably go into creating and distributing your promotional materials. Promotion is the conveying of information about your business and its services. Some of the ideas that are used today include; sales letters to appeal to a customer that is considering using your services, brochures, a simple piece of paper folded giving information about your services, product sheets, a description of your services coupled with quotes from satisfied customers, promotional kit, a simple folder containing all of the items mentioned, perhaps with some pictures of a very well done job.

Most businesses fail because the owner does not know how to reach their customers. Marketing is one of the ways that you can get people who need your product or service to find you. It is one of the main reasons some businesses charge a lot for their services, and never do a good job, or have good customer services, but are never busy enough to pay the bills. They make themselves available to their customer base and when that customer needs a service they are always easy to find.

The struggling business owner typically provides a better service at a lower price, but is unable to get enough work to survive. What separates a successful business from an unsuccessful one are the marketing efforts to their potential customer base. Think about how many things you have bought that were not near the quality of what you expected them to be. I bet all of them had one thing in common and that was a good marketing campaign.

Let your potential customers know who you are before someone else does. Also, sharpen your skills and become an expert in welding, this can be a very potent selling tool. It will be a very important part of your brand. Make your name synonymous with great work and knowledge. Make your customer base so impressed with your work that they feel that they are the only customer that you have. Your best customers are testimonials to your work. Let your future customers know that these customers use your services. This demonstrates that your customers are important to you. The customers you spotlight can also benefit from your marketing. Utilize customer centered one-to-one marketing; it is a very important part of your business.

How to Market in Print Items

In print items like business cards, flyers and general promotional materials are a good way to get potential customers to know who you are. It is the start to establishing your business image. This is a long process that is combined with networking later on to establish a personal connection. The key to this type of marketing is you need to do it consistently and you need to do lots of it in different places. It is by no means a guaranteed way to get work, but more of a way to establish your image before you meet those people. Make some flyers and promotional materials then mail them out to your potential customers. Even better yet go out and hand them out yourself. Business owners have a lot of respect for other business owners who are not scared to promote themselves simply because they had to do the same thing to get to where they are now.

When it comes to promotional materials, create several types, targeting different types of customers. For example make materials for restaurant equipment repair services, onsite heavy equipment repair and a final one for automotive parts welding. The materials need to be targeted to that type of welding. You want the people receiving them to think that you are an expert in that area of welding even if you have never done that type of work. Mail them out to each type of business regularly.

The restaurant repair one would be sent to all local restaurants and restaurant equipment suppliers, the heavy equipment repairs flyer to every heavy equipment contractor, truck supply store and shop and finally the automotive parts welding service to every automotive repair shop in your area. Do these mailings once a month for three months and then once every three months from that point on. After a while these business owners will begin calling and if you decide to go meet them and introduce yourself they will already know who you are. That is how print marketing is done. It is a long process that typically pays off in six months and is combined with networking. It does work, but you need to stick to the plan.

It must be mentioned that there are many books on marketing that state if you send out post cards and flyers alone you will get business in the door. Yes, that can happen but it is not very likely unless you keep doing it for a year or more and follow up with a phone call in six months to everyone who received your post cards. You have to wonder why some of these books state this. The catch is they were written by printing companies and their ultimate goal is to get most of your money spent on printing. They offer a false hope for the many business owners who dread cold calling or knocking on doors. Be smart, commit to a plan before you start.

Business Ads and Sponsoring Events

Business ads are a good way to get new business in the door only if they are put in places where people who need your service are looking. Newspapers and magazines are not a good venue for advertising your welding business. You need to find industry publications that are related to the metals industries or products, but do not have many welders advertising in them. Ideally, you want to be the only welder advertising in the publication or on a sign. The trick is to find niches where your services are needed.

For example advertising in a local farm equipment newspaper might be a good place to get repair work. Another place could be in a local racing circuit or track where people need emergency repairs. There is no simple answer to what will work but I will say most types of advertising do not work and only benefit the sales person who sells the ad space. Ads are not cheap and you must find the right medium in which to advertise. There are three things to think about before advertising.

1. Do the people reading the publication need what I have to offer?

2. Can these people afford what I have to sell?

3. Am I the only welding business advertising?

If the answer to all of three questions is yes, then, you must determine if this is worth your time and money. If the answer is yes, then you must commit for at least six months. It is like spending on printed materials. People need to get used to you at first, then some trust is established and finally, and most importantly, your ad needs to be in the right place at the right time.

Sponsoring events is a good idea for some very specific industries. If you attempt this, do it in a way to network with people who may become customers or even refer business to you later. Do not sponsor welding or fabricating related events. These people are your competitors, and are after your work. More than anything they will be trying to sell you something. If you sponsor an event or have a booth at one it needs to be in an industry where you are needed just enough that the current business owners do not need to hire a full time welder. An example of an event that might be worth attending, sponsoring or getting a booth at are home shows. There will be lots of contractors, architects and engineering services that once in a while need welders to make gates, fences, iron work, railings, staircases, etc. They have a need for your services, but not a great enough need to justify hiring a full time welder to do the work in-house. The only exception to this is if you are promoting a unique product and the event is full of potential buyers. An example of this would be you have a line of marine accessories and decide to attend boat shows to showcase your products to the public, dealers and wholesalers.

As a start up, stay away from events and advertising in general newspapers. If you do decide to do that, then do it in a very industry specific publication or event. Some examples of were to advertise, attend or set-up a booth are:

- Boat shows to showcase marine accessories or meet potential customers.
- Home improvement shows to meet general contractors, architects, designers or to show your products.
- Advertise in local restaurant supply newspapers or magazines.
- Advertise in industry specific manufacture newsletters.

Websites, Online Marketing and What to Expect

Websites are a great tool for marketing and establishing trust. What they are not good for is getting new business in the door. Websites are difficult to locate in the search engines and advertising them can be very costly. Most people who call a website's phone number are looking for either emergency repairs during odd hours or have some small item that needs a weld and they can't afford much. Either way, these are not the types of customers that are going to pay your bills.

What a website is good for is to establish trust with your potential customers and it is a great way for them to learn more about what your company does. Your website can be a big part of creating your company's image, and there is a very small possibility that it might also bring in a new customer now and again.

Should You Have a Website?

The answer is yes! Having a website available to your customers 24 hours a day is extremely important. A website is the new business card/flyer/billboard. More and more people are turning to the web for information and to research a business. If you have a site, then that is your way of showcasing your work with pictures, videos and your qualifications. It may not be the best way to get work in the first place, but it is a great way to close the deal and get the trust needed to get the big contracts signed.

When designing your business cards, logo and promotional material you should also have a website built. Think about this; you are looking to buy a new car and you typically start your search online. Chances are that you want to see pictures and videos of that car. The sites that give you the most information are likely the dealers you will call, and visit later on. Does this sound about right? That is exactly how your customer is going to use your website. It is just another way to build the trust and recognition you need so that you can close the deal. Better yet, if the sales person has a picture or video of them, then you already feel like you know them, or at least know who you are dealing with. These same rules apply to a welding business. People buy from people, or products, they know something about, and the more they learn about you, the more likely it is to get the business in the door and that contract signed!

Building a Website

If you are going to have a website then you should have someone build it for you. Better yet, don't go with a custom website but a proven template (these are a *lot* cheaper and better). The things that should be on your website are:

- Business Name and Logo
- Location, Address and Service Areas
- Contact Information with Phone Number
- Types of Work You Do

- Lots of Pictures, Videos Information About Previous Jobs
- Pictures of you and your family, your truck, shop or anything to let people know who you are.
- Guess what folks? You know this stuff works because you have also fallen for this these marketing techniques.

Before starting a site you need a domain name. A good piece of advice is to choose a name that has the word 'welding' in it and that is 'brandable' or catchy. For example:

www.xyzweldingservices.com

www.mycountywelding.com

www.aluminumalswelding.com (Aluminum Al's Welding Services).

You want to make sure you include your company name, or at least the location, in the domain name as well as the word 'welding' for later marketing efforts. If you are opening a manufacturing business, then you'll want either just your business name or product name in the domain name.

A basic rule to go by is *NOT* to buy a long domain name just because it has all of the information you want in it. That is a whole other subject so just take my advice on that one. If possible get the ".com" and don't let them sucker you with the need to buy the other extensions like

.net or .org. These have nothing to do with people finding your website. The websites that are the easiest found in search engines have the most incoming links that are from trusted sites on that subject. The domain name extensions are just another money making scam that prey on new business owners.

As for the subject of websites and web design, it is slightly complicated, and beyond what most welders are comfortable learning. My advice is ***not*** to make it a do-it-yourself project; hire someone to do the work for you. It is a good idea to learn the basics of how to build websites so that you can later make changes yourself by adding pictures, videos or whatever you want.

If you don't know how to make your own changes, be aware that you will likely be paying approximately $60.00 or more every single time you want to copy and paste a single sentence, insert or move an image, add a page, etc.   You could easily learn to do this yourself and make all your own changes for free, not to mention more quickly than they could. Not only will they charge you these outrageous amounts, but they also like to make a big deal out of each request! It's simply not worth it to pay someone else for this kind of easy fix!

Here are some places where you can get domain names, locate companies that build websites cheaply, and free web design classes and training:

Domain names, websites templates and more.

www.GoDaddy.com

www.NetworkSolutions.com

www.Register.com

Web Design Services

www.ABCWebsService.com (full service with excellent customer support, pricing and highly recommended) Mention promo code "WeldingBiz" from this book and you will get discounted pricing without sacrificing service and quality.

www.GNC-Web-Creations.com (free and paid web services)

www.LogoNerds.com

Free Web Design Classes and Training

www.CodeCademy.com

www.W3Schools.com

www.htmlDog.com

Online Marketing

Online marketing comes in three forms:

- PPC or Pay Per Click Advertising
- Online Ads
- Organic Search Results

Pay-Per-Click Advertising

PPC or pay-per-click advertising works by you bidding on keywords and phrases that people are searching for. If you win the bid, then your ad is shown to those people searching for those terms. It can be set to target areas like zip codes, radius from a zip code, state, country or anything you can think of. If you win the auction and your ad shows, then you only pay if someone clicks on your ad and visits your website. For example you would bid $5 per click for the term "my county farm welding services" for a 25 mile radius from your zip code. What you are searching for is someone who needs their farm equipment repaired by welding. It is a good system if you find the right combination of key words and locations to bid on. You can find out a lot more about this type of advertising through the following programs:

www.BingAds-MicroSoft.com
www.Advertising.Yahoo.com
www.Google.com/AdSense/ (The most traffic and most expensive but has the greatest potential to find customers)

Online Ads

Online ads are just like print ads. They only work if you find the right place to advertise and have little to no competition. In most cases, these ads will lose money and only benefit the publication owners and the sales person who sold you the ad space. There are some exceptions, however, specifically listing your site and business in the niche directories. Some example of niche directories that *may* work are (the keyword is *may* because there is no guarantee they are worth a penny to your business):

www.Yellowpages.AOL.com

www.Local.Yahoo.com

www.Local.BOTW.org

www.Thomasnet.com

Organic Search Results

Organic search results are the results that the search engines show without having paid to be there. The term used for this type of online marketing is SEO or Search Engine Optimism (Google the term "SEO" to learn more about it). These results can take years and the marketing techniques used in this area require very advanced marketing and networking skills. The worst part is that most of the information out there is wrong, or even dangerous enough to get your site banned from the search engines. I do also want to say that most companies that offer SEO services do not know what they are doing. Much of the misinformation is coming for some of the biggest brands praying on new website owners. If they did know something of value they would not be sharing that information with you because they would be working on their own site. What you need to know is this:

- Make sure your website lists the types of services you offer.
- Have your location, service areas, address and phone number listed on the site.
- Have some content that is written about what you do.
- Label and title your pictures accurately.
- Get your Meta Tags labeled properly including Title and Description with the proper H1 through H4 Tags (The free website building classes cover this)
- Links are what build trust and how you build traffic so make certain you get links to your site.

Before any of this is done you want your site to be built static and stay away from java script as much as possible because it does not do well in the search engines. The more HTML code you have the better.

SEO comes down to matching the title, description and content of your website so that it all tells the same story to the search engines. It is like choosing a book. First you read the title and then look through the list of chapters in the table of contents. Finally, you read the inside of the book or content. If everything matches, you are happy. Search engines want to see titles, descriptions and content matching. If not, there is something wrong and you get a penalty. Links coming to your site are like votes that tell the search engine if your site is what the title and content say it is. Links that come with keywords to your site are how the search engines will ultimately rank your site. If you get links with keywords that say welding, my location, types of services then that is how the search engines will place your site.

For example, Bob's Stainless Steel Welding Services, in My Town USA. You want all of your content showing this information and different links containing these key words to point to it. If it is all mixed up properly then you will get visitors based on those search terms. If you have no clue what you have just read actually means… then don't worry about it because it is not worth the effort.

If you do decide to take a try at marketing this way, then you can visit Google Webmaster Central to get more information. It is free and well-worth reading about marketing your site.

Online and In Print Marketing Services

www.TankMarketing.Com (excellent company that I personally use)

Online Marketing Classes and Information

Finally here is a list of worthwhile places to learn from that are free or at least worth looking into to find someone to do this work for you.

www.Google.com/Webmasters/ (Direct information from Google)

www.GNC-Web-Creations.com (offers free online classes and excellent services)

www.SEOBook.com (Advanced high end and a great place to hire some up-and-coming talent for competitive terms)

www.SEOMoz.com

www.SearchEngineLand.com

Here are some places where you can build some general links and learn more about writing and article marketing.

www.EzineArticles.com (Write articles for links)

www.Blogger.com (Start a blog)

www.Squidoo.com (create a page with links)

Find industry specific websites and write for them in exchange for a link from your article.

Press Releases

When you are starting any business, a press release is definitely a very good idea. How well a press release does all depends on how unique your business is. For a general welding business a press release is not a good idea unless you are using it to promote your website. In this case, it is a form of getting links to your site and a way to tell the world about it. Also, a press release is a great idea if you are launching a specialty service or product. It is a way to get the media interested in your product or service. In some cases you can get lots of media coverage, and that is priceless in terms of launching a product. Here is a list of press release services.

www.PRWeb.com

www.PRNewsWeb.com

www.BusinessWire.com

The Truth about Social Media Marketing

Social media marketing has no real effect on your website, business nor customers because they really don't care unless you have something out of the ordinary to offer. For most businesses the value of social media sites like Facebook, Twitter, Youtube and Linkedin is that you get a link out of it to your website. Since anyone can get these accounts the value of that link is pretty worthless unless your website has no links whatsoever. That is the truth and please don't buy into the hype that this type of marketing is the way to go.

On the other hand, you can hit it big if you use this type of social marketing to 'shock the world' with the media coverage of the fire breathing dragons from old cars that you are fabricating, if you get my drift. But be warned that social media sites are full of millions of people hoping to hit it big like the lottery. As a welding business the social media marketing is more of a novelty than anything else. It is good way to get seen, but don't expect much or anything from it except that your customers might have more trust in you simple because they can find you online in other places.

The final word on social media marketing is that it can work for the *right* business or product. For example, if you are building the most awesome custom bikes/choppers on the market today, then that would work. Bikers would be lining up to buy your choppers and business would be awesome because of a video. As a regular welding business it would be worth posting on some social media sites like Youtube. Simply to introduce yourself (include your family because people love that), show your skills and put that video on your website so that you can build more trust with the people you are marketing to. Social media is nothing but another tool to build trust with your customers. It will not close any deals or get you any work in the door! Social media lets other potential buyers learn who you are and decide if they want to do business with you. Hint, it is basically a free commercial and ad space for your business. Everything else is done in person. Just remember, people buy from people and not companies, websites or marketing material!

Social Media Sites

Here is a list of social media sites you should have a profile or presence on:

www.FaceBook.com

www.Twitter.com

www.LinkedIn.com

www.YouTube.com (Film an intro video of you and your work and post it on your site)

www.FlickR.com (Post pictures of your work to share)

Why Losing Money Is Good

Losing money is a good sign when starting or growing a business. Specifically, losing money on the right investments is a good sign. If you are sending out flyers, paying for ads, spending a lot on gas because you are meeting potential customers, then that is a good thing. Running a business is a numbers game. If you make so many contacts, send and hand out so many flyers and business cards, then there is a number of customers that you will get from that work.

For example if you meet 50 people a week who might have a need for what you have to offer and it cost you $1,000 dollars to meet them. Now if that marketing effort results in you getting $10,000 worth of work Now you know exactly what it costs your business to get $10,000 worth of work. Let's say your goal is to earn a gross income of $20,000 months, then you know it will cost you $2000 in advertising and you need to meet 100 new people who might need what you have to offer to get to that income level. Make sense?

As a business you need to know what it costs to acquire a new customer. If you know the cost, then you know how much time and money you need to spend to make what you want to earn. It is not uncommon to spend 25% of your income and 90% of your time on marketing and networking. It is all worth it as long as you keep growing your business and income.

Have you ever wondered why banks keep buying other banks? It is because it costs less than advertising and getting a new customer in the door. They simply buy the bank and adopt all of their already existing customers! That is exactly how you need to look at it. If you spend $5 dollars a click for a website visitor and it takes 100 clicks to get a job, then as long as the job pays enough, it is well worth spending that $500 to keep getting more customers.

You will not have enough business to survive if you are not losing money to marketing and networking efforts. You need to spend money to make money. For most businesses there will be a time they get in trouble because they either stop spending on marketing or cannot afford to spend anymore. If that happens to you, you're out of business.

Analyzing Marketing Cost versus Your Return on Investment (Marketing ROI)

Owning a business is no different than investing, except that you have more control over your own success. When you invest in stocks, bonds or whatever, you are at the mercy of the management team's decisions. When you own a business you decide what costs are worth it and which ones are not. As a business owner you have full control of a company and can do whatever you want with it. You can run it into the ground or make is the next fortune 500 company like GE, Caterpillar or even Fluor.

Those businesses started exactly where you are right now. A dollar and a dream. Willing to do what it takes to succeed. The successful ones analyzed their marketing costs and figured out the ROI, or Return On Investment, they wanted to achieve. They planed ahead and stuck to the plan. As a new business owner you need to find out what it costs to get each customer in the door and how much money you will make on them. These are all averages and no customer or job is exactly alike unless you are producing a product and control the price. The bottom line is the bottom line, and you need to figure out what you are willing to risk money wise to get that business.

ROI is the way you decide how to spend your money. Knowing this information will help you determine how much you should be spending and what you can reasonably expect in return. Most welding business owners have no clue and spend lots of time and money but never get anything in return. You need to play these numbers right and if you do, you can build an empire simply by investing your time and money properly. Later on you can expand on them by having other people do your work for you.

If you are making $10,000 for every $2,500 you spend, then your ROI is 400% on your marketing investment. These numbers are important because they let you know what your budget is and what you can risk.

CHAPTER 6 - NETWORKING FOR CONTRACTS THE ULTIMATE MARKETING TOOL

Networking 101 - Welding as an Employee vs. Welding as a Business Owner

One of the best free resources for welding business owners is networking. This may be done at any event or situation where there are people. Almost anyone that you meet could be a potential customer, or might be a contact to a potential customer. When speaking with people, tell them your interests, and most importantly, listen to their interests as this may result in finding the contacts that will move your business toward success.

It is vital to make an effort to meet new people. It may be that one person who will be the key to large contracts. To do this, you must first show interest in the person, and what they do. That is an introduction to bringing the conversation around to your business, and your interests. Hundreds of welders provide similar services at similar pricing but the question is how does a customer decide on the welder that receives the contract? The answer is relationships.

Customers will give business to companies with which they have a relationship. Certain questions must be answered. Is the welder committed to the customer? Has the welder provided the best service? Has the welder kept the relationship, or just fulfilled the contract and went on to other things? Remember, you want to develop and establish a trusting relationship between yourself and the customer.

Networking goals:

- Work to build a long-term relationship with the potential customer.

- Always be selling the benefits of your business.

- Provide wonderful customer service.

- Accomplish a full commitment to the customer.

- Keep selling and always be closing.

Networking is an area that most welders do not think about. How willing are you to network and sell yourself to get a welding job? Chances are when you need work, you get out there and start calling numbers and sell yourself so you can get a weld test scheduled for the job.

As a business owner you might be fearful to call anyone, meet in person or send out marketing material to push your services. The point is, if you need work, then you do what needs to be done to get that job. Many welding business owners wait for work to find them, and it never does. Unfortunately, when the work does not magically show up at their doorstep, most welders will simply change gears and end up going to work for someone else in the end. Their dreams of owning a successful business and working for themselves are shattered.

What has happened here? Why would they be willing to work so hard for a low paying welding job all to help make someone else rich? Networking would have made it possible to work for their own benefit. Does this make sense, and why are so many welding business owners doing this? It is the fear of selling yourself for your own business.

Here we come full circle back to the first chapter of this book in which the question was asked, " Are you ready and able to do what it takes to start a welding business?" This commitment is **not** for the majority of welders out there. You must figure out if you wish to be a business owner, or an employee. It all comes down to what makes you happy. Business owners are never happy as employees, and employees are never willing to do what a business owner needs to do or is always complaining about the work required of them. Where do you fit in?

Relationships and How They Rule Our Lives

Do you want to know how real business works? It is all about relationships. Have you ever wondered why some CEO's get multimillion dollar salaries, perhaps do a bad job by running the business into bankruptcy and don't even show up for work more than once a month? If it were any other job they would be fired on the first day. The truth is that the investors and the board of directors *like* the person, and even though they are making a mess of things they still get that salary, power and always have a job. They know how relationships work and know how to take advantage of that situation.

Most text books say if you work hard you will be rewarded. Even though that is mostly true, it is not exactly correct. In most business situations there is a group of people pulling the weight of the entire company with one or two people taking the credit for their work. The owners and management teams have become experts at using relationships to their advantage. That is life, and it is not fair to those who do their part, but that is how business works.

The power of relationships is what most CEO's have mastered. Yes, some are business savvy, make smart business decisions but they are being paid for who they know and their ability to close deals. Think about this, how many CEO's spend their days at charities, playing golf and socializing with other CEO's at the country club? You might think that is a waste of the company's time and money but that is how they build relationships and trust with other businesses. This all leads to making deals, mergers, and earning the trust of the most influential people in the world. This group would include; other CEO's, politicians, lobbyists, and generally movers and shakers who make the big decisions. This all leads to signing contracts that keep their company busy with work and it gives them the inside information and influence it takes to win lucrative contracts and jobs. There is no other way that these jobs could be earned, except by knowing the right people. Those "right" people are the ones that make the decisions as to who will get that multi-million dollar contract.

In the world of business, or life in general, those who are well-liked, good-looking, or at least play the part well get all of the benefits with the least amount of work. The people who understand the power of relationships know they do not need to do the job correctly because they can get someone else to do that for them, or at least fix their mess. They just need to make the connections because they know people will like them and sooner or later they will use that to their advantage.

It is the same basis of any con artist scam, and it works. We want to trust people whom we like, and perhaps, who could do something for us. It is an undeniable attraction that blinds us from what someone's true intentions are. In the world of business, this is considered a tactical edge the same way a position of high ground would benefit a sniper.

Let's push the relationship subject farther. Suppose a stranger were to ask you for $500 to repair their car, what would you say? Most likely your answer would be no and you are out of your mind, I don't know you. Let's flip the relationship status to someone you know. Suppose your mother asked for the same $500 for auto repairs. You would most likely give her that money with no questions asked because you would be proud to help your mother in that situation.

That is the power of relationships and in the business world it is an edge to get what you want. This is why marketing and meeting people work hand in hand to build trust. Build trust and ask for the deal later when they cannot refuse. The more trust you have the less you need to work to get the job or contract. It is just human nature and nothing more. You must learn to take advantage of this human nature without feeling guilty for doing so if you are to succeed.

Why Networking is the Ultimate Tool and the Core of Any Welding Business

The world is one big social circle and as a business owner you need to belong to the right circle. As a welder this is also true. Most of the high paying traveling jobs go to the welders who know a foreman or recruiter. As a new welder you start chasing lower paying jobs advertised in papers and magazines then eventually you make some friends. We all talk, because we need each other, and perhaps help each other. We find out who is hiring, and we travel in groups. Later on, we meet the right recruiters and foreman. Before we know it we have all of the work we need or want, but then start picking the jobs we want because people are lined up to hire us. That is exactly how the business world works. You need to start meeting the right people who can do something for you and are able to give you the work that pays the bills and hopefully more. This is done in person because people do business with people they like. If you cannot, or will not do that then forget about running a business, because you will fail.

<u>If your personality is not a business owner's personality then you need to accept the fact that it is not for you, or you have to find ways to re-shape your personality to make you a more outgoing, sociable person. There are books and even courses available to help you break out of your shell for this purpose.</u> The reality of running a business is that it is not for the majority of people. If you learned one thing from this book and that is this lesson, then consider the money you shelled out for this book well spent. Just be glad you did not sink your life savings into a welding business, as many already have, only to see it all slip through your fingers simply because of your natural personal style. You will have avoided a personal disaster.

In the end, you will need to market your business in order to let people know about your company. This builds some trust over time but it is a stepping stone to when you go out and meet the business owners, politicians, engineers and the people you need to meet. You want your marketing material known so that when you knock on their door, they say "Oh, yeah, we got your post cards, and were going to call you for a job". Since you are there now you are ready to get that job. That is how it always works. They never call until you meet them in person and then there is lots of work. You will become a person-to-person seller of your services. This is another word for direct sales. Direct selling is the sale of your services in a face to face manner. You will meet and socialize with potential customers. Your bottom line is directly related to your selling. You are the owner of your company, in order to survive; you must make sales calls on potential customers.

The People You Need To Avoid Who Steal Your Time and Money

Want to lose money faster than you can make it? Then hang out with this crowd:

- Welders
- Fabricators
- Welding Supply Store Owners and Visitors
- Friends and Family

The point of this list is that you need to focus your time and energy on meeting the people who need your services and might pay your bills. This is another area where most business owners go wrong in managing their time and end up working for someone else in the end.

Other welders are not going to help you get work because they want it themselves. Besides that, they also want to own their own business. Fabricators are also looking for the same work you are. Welding supply stores are just looking to sell you more junk, and probably have a friend or family member to whom they already refer any work that comes in. Finally, friends and family are well meaning but you need to devote time to your business.

You must ask yourself, if I associate with these people, is it in their best interest or mine? Chances are they need you and you don't need them, and that is why they associate with you. Not a good business decision.

The People You Need to Meet Who Pay Your Bills and Need You

Think about this? World leaders associate with other world leaders. CEO's with other CEO's for the same reason. The poor with the poor for the wrong reasons and the rich with the rich for the right reasons. The saying "birds of a feather flock together" is so true. You need to find the flock that is good for your business. Who are the people who can make your business grow and most importantly need you? Here is a short list for starters:

- Engineers
- Architects
- General Contractors
- Restaurants Owners
- Machine shops
- Car Repair Shops

- Performance Shops
- Boat Dealers
- Engine Repair Shops
- Heavy Equipment Rental Businesses
- Heavy Equipment Service Contractors
- City, State and other Local Engineers
- Politicians in Charge of Public Works Projects
- Plant Superintends
- Building Superintends
- Marinas
- Dock Masters
- Project Managers
- Small Airports
- Aircraft Parts Remanufacturing Shops
- Manufacturing Facilities and Factories
- Crane Services
- Corporations, LLC's and Organizations that Outsource Welding and Fabrication
- Purchasing Departments
- Outsourcing Services
- Anybody who deals with metal and is not a welder, but does not have enough welding to justify hiring a welder in house.

This is just a basic list but it is a list of the types of people who might need welding services. Most won't give you a job, but it is a lesson in thinking about where you can get work from and who might need you. The choices you make as to who you associate with are going to influence your business a great deal. Make the right choices and choose people who need you, your product(s) or services.

Getting the Job Done with Incentives

Want to get decision makers to give you lots of work? Throw in an incentive. Other words for an incentive include coupons, discounts or referral fees. This is the business world and that is how it works! Cell phone companies want your friends and family signed up, while others are looking for your general commitment. Incentives are hidden all over the place and businesses have been using them for years. Cable companies are battling satellite companies for customers and all of them are using incentives to get the customers in the door! Does this sound familiar "Get $50 just for signing up"? Do you think that business is giving away $50? Obviously not and they know by giving you $50 to sign up it will cost less than marketing to get someone new to use their service. In the end it is a war of pricing and loyalty but not services. In most cases, the service is identical.

As a business owner you need to create incentives where other businesses are at the mercy of what you have to offer! To start, advertise "Free Estimates"! Think about what kind of coupons you can offer, the types of discounts you can give and what you are willing to do to get the business in the door. What can you do that nobody else is willing to do for the price if they buy enough of your services.

The key to offering successful incentives, is to give away a little and in return the customer spends a lot more. It is preying on the cheap and greedy business customer base and using that cheapness and greed to your advantage. Hint, the free lunch (never free) that costs less than to get a timeshare buyer in the door otherwise (ROI or Return on Investment). The goal is to give them just enough rope to hang themselves by signing the contract, but nothing more.

This is a time to think and get creative. What incentives have you fallen for in the last few weeks or years? Two-for-one, buy-one-get-one-free, refer-a-friend to get $50 off your next purchase, buy nine and get the tenth free and so on. We all love a great deal and getting that feeling that we just made off with a great bargain. As a business owner and a consumer you need to understand the power of incentives and what they can do for your business.

In the end, coupons, discounts and incentives are great for establishing a new business. They are by no means a good way to grow your business. It is nothing more than a way to hook people to get a taste of what you have to offer. Hopefully they will like you as a person and pay more for your services down the road. Incentives work like a charm and are better than working for someone else. It all depends on your future goals and what you are willing to do for the work. Here are some incentive ideas:

- Free Estimates
- Referral Fees
- Grand Opening Discount
- Volume Price Breaks
- Hourly Price Breaks

Free estimates are one of the most basic incentives that you should make sure to use. They give you a chance to meet potential buyers and then you can focus on how to earn their business. Many buyers will base their decision on the person they liked the most and not the price. The free estimate is a way to build some trust in yourself as the new kid on the block and that *is* what closes deals.

Referral fees work great when you offer them to the right people. This is an area you do need to be careful in because it can sometimes fall under the term 'bribery' and land you in trouble. An example of a referral fee would be to approach a Dock Master or even a Handy Man and offer them $500 for any job that is over $5000 that they refer to you. To some people that $500 will motivate them to try to find you work. The dangerous part of referral fees is knowing who you can or can't offer them to.

Approaching a politician and offering a kick back for a contract could land you in jail. On the other hand, if you could afford it, you could hire a lobbyist for a fee and they can get that same politician to give you the contract. Another way to offer a referral fee is to hire a sales person that works on commission. They go out in the field, meet people and close the deals. They do not need to know a thing about welding because they are in the business of forming a relationship.

Be aware that if you hire a good sales person they might just take your work with them and open up their own company by hiring a welder to do the actual work. In this case you need to get them to sign a non-compete agreement before hiring them on that prevents this from happening.

When opening up a business you need to make a big deal of the grand opening and offer some type of insane discount. This will pay off in the long run because people will know about your business. It will also bring in those very cheap business owners looking for a cheaper welding service to use. Many contractors spend their lives shopping for the cheapest service, and don't care about anything else other than the lowest price.

Volume and hourly price breaks work great on the cheap customer and are a good way to establish a relationship. They are good when you are starting out but be warned that this is no way to run a business for the long haul.

How Successful Welding Businesses Market and Sell Their Services and Products

Welcome to the world of big business. You are nobody and don't stand a chance to play at their level. How do they do it and make so much money with so little effort? Simple. They invest wisely and calculate what is worth spending money on and what to stay away from.

The rich and successful don't waste a moment when it comes to profit. They do what needs to be done and have no problems spending as much as needed to get the contract signed. They can spend these amounts because they have earned their share, know the market and are calculating risk or ROI. As a small business owner you focus on cutting costs and figuring out ways to stay afloat. Most small business owners are successful because of their ability to do jobs cheaper or faster.

The rich and successful welding and manufacturing businesses focus on making more money. They focus on making more money instead of cutting costs. They want to lose lots of money as long as they make it back many times over. The more you lose today the more you will make tomorrow. It is an investment to them and nothing more.

Think of it this way, if you can make $5 for every $1 invested, would you keep investing more and more or only do it a few times to make just enough to pay your bills? Chances are you would do it and make as much as you can. Successful big businesses focus on ROI or return on investment. This is what separates the small business owner from the most successful welding and fabrication businesses.

As a small business owner you struggle to stay in

business, and your instincts are to cut costs. It works in the beginning, but if you want to expand, it is a plan for failure. Employees get paid the bare minimum, supplies are shopped and chosen on cost and jobs are won by being the cheapest. This all leads to getting the least skilled employees, the bare minimum in material quality and a lack of service needed from all of the people to let your business grow. Not a good combination for any business unless you can change that over time. Cutting costs works, and is a must while starting out, but when you are competing for the big contracts these strategies will land you in big trouble.

Big business does not go after mom and pop deals. They go after deals that are no longer available to the small business owners (aka: mom and pop retail shops). This means more profit, less competition and more control. It is a formula that allows for big mistakes because the profit margins are so large. Jobs are won by bidding right, knowing the right person or paying salesperson/lobbyist to get the contract. They consider those expenses a CODB, or the Cost Of Doing Business. The successful think of a CODB as a logical expense.

Here is an inside tip on how big businesses use their time and money wisely. If they are bidding on jobs throughout the state or region but don't have the manpower to have a person sit in on a meeting, then they hire a private investigator to do it for them. For about $1500 you can have a Private Investigator sit in on your company's behalf and take notes, photos or whatever is need for you to submit a bid for a job. If they spend $1500 per job and it takes 20 bids to get the job. Then spending $30,000 on private investigators pays off if you get a large enough contract.

This again is taking into account your ROI and losing money for the right reasons. Who cares what you spent as long as you make that money back many times over? This is a common business practice among mid-sized businesses. Once you grow out of the small business stage you need to learn that it is no longer about costs, time or working hard. It comes down to making the right connections that can do something for your business. You need money to get the right insurance, bonds and pay the right people to do the work needed to get the return on your investment. It is all about getting the right people to do your work and focusing your time and money on getting more business.

CHAPTER 7 - GETTING YOUR FOOT IN THE DOOR

How Are You Going To Plan to Make It Happen

Think about this, incentives, marketing, overcoming competition, and getting government contracts.

The early door-to-door salesman, who did not wish to have the door slammed in his face as he tried to close the customer, actually put his foot in the door to prevent its closing. How do you and your new business figuratively get your foot in the door?

If you decide to own your own business, you may think that you are going to buy a welding rig, open a shop and work will just flow in. If that is your plan, you are in denial as to how businesses succeed. You will end up working for someone else if you don't have a clear, logical, financially sound business plan.

What are the *individual* steps you are planning to take in order to make your dream of owning your own business come true? The *precise* steps? How *exactly* are you going to make it happen? What do you have to offer that no other welding shop or business can do? This is another thinking game/exercise that you need to play. Keep in your mind; "fail to plan, plan to fail". It is better you think about this before you put money behind your bets. It is a fact of life that people who have no ideas about their financial situation or just get by without planning will fail the majority of the time. Develop a game plan within your financial and physical means, and figure out how you will make it happen.

The $20 Ball Point Pen

Think about this. One of the most common mistakes all of these businesses do is buy a $20 ball point pen for their employees. The actual pen only costs ten cents but management does not like to spend money on pens or anything unnecessary. So the employee ends up spending about $20 of company time looking for a pen to simply fill out a form or put their initials and name on a time card. Better yet this $20 pen is the gift that keeps on giving to the employee. The employee goes through this almost every day.

You see, the pen is actually cheap but management is blinded by the cost of pens and fed up with how many never come back. If you tried selling them that same ball point pen for $20 they would be outraged, but they have no issue spending that much on a pen in other ways. That cheapness and blindness for a bargain is irresistible to most business owners and management in general.

As a business you need to create incentives that take advantage of that weakness (cheap management and decision makers). They are their own worst enemy and all it takes is the right deal and you have them hooked. Just like the $20 ball point pen, you need to find a different way to get your money out of the customer. You might not get the $20 for selling a pen but you can get it in time wasted while on the clock.

Incentives for Companies

Your customers hate spending money because the money they spend hiring you, is less money they will take home themselves. Most businesses are in the small-business-owner-mode. They tend to only figure out what they are paying out, but not what they get in return. For most successful small business owners they would rather lose a high paying job than allow anyone to make any money on them. They are blinded by the dollar figure and have no clue about their return on their investment (that pesky ROI again). This is common among big and small businesses.

To understand why you need to offer incentives you need to know your buyer. Most of the buyers of welding services or metal products are concerned with two numbers:

- Hourly Rate

- Project Cost/Product Cost

Honestly, the customers do not care about anything else. Sure, 'quality' of the work or product sometimes comes into play, but nine times out of ten of them only care about their bottom line. It is a simple answer that they want and will pay a lot more money if you know how to sell them on the cost of the work or product! No matter what you have to offer, these two numbers will rule the buyer's decision and consciousness.

What you or your negotiator/sales person needs to do is figure out what the main issue, or what their "hot button" is! Hint, the first objection is the last objection and the deal closer (pay attention here)! When you meet with the potential buyer the first words out of their mouth are going to be the main obstacle you need to overcome. If it is the hourly rate issue, then give them an hourly incentive. This way they will hire you for many more hours simply because they want the discount (think 'buy nine hours in advance and get the tenth free'). On the other hand, you could pitch a project discount to those customers that are a bit more sophisticated. Maybe they will agree to spend $25,000 or more in order to receive a 10% discount on the overall price. This way, they will give you more work than they need until you get the job done or have earned back your investment.

You must address the client's primary concern first! If you do not address their biggest doubt immediately, you will lose the work to someone else who will. You need to make them feel that they got a bargain right away and close the deal before they get a chance to shop around. Use urgency and tell the customer that you are slow for the next few days, but after that you have work lined up for a few weeks and won't be able to give them that deal any longer. Also, be very prepared to get rid of a client that wastes too much of your time negotiating because they are going to be an emotional and financial drain on your business. Here are some ideas and suggestions:

- Only NEW customers are eligible to pay for four hours of service and get the fifth hour free.
- Buy two ladders and get the third free.
- Fabulous Friday Fabrication Day – Everyone receives 10% off Fabrication Jobs.
- Refer a friend and get 15% off of the next job.
- Free installation with the purchase of $1,500 or more in fabrication work.
- Free Delivery.
- Free Anything.

You need to be creative and find ways to hook the cost conscious business owner. Pricing is what most buyers can't seem to overcome so you just need to find a way to make it irresistible to them. You need to create the game for them to get emotionally involved in. Once you've got them hooked, they're yours! Just look around at all of the incentives that are surround you daily everywhere you go.

Business is War and Conquering Your Competition

In any industry there is a limited amount of work. It so happens that 80% of the work goes to 20% of the businesses. Successful businesses get the majority of the work because they find ways to earn it from their competitor's customers. You may be able to get some of those contracts that current customers are not happy with, if you plan properly. When you market and go out to meet potential buyers or customers you need to listen carefully to what they are saying about your competitors. You must learn to quickly address their concerns by giving them a solution that can come only from them hiring your company. If it is price, under bid their current welding shop, if service, offer them more in return for the job. Do whatever it takes!

Here is a secret about most established businesses. Once they become successful, many owners become lazy and careless. This is a perfect time to step in and take that work away from them. Keep in mind that these potential clients are also business owners that love and will gladly take advantage of a new business in town that is willing to do what it takes to keep them happy. They have all been new business owners once before, and they are more than happy to give a new business a chance. However, be warned! Stay away from trying to steal work that was given out to a friend or a family member. Don't waste your time. This type of relationship is too strong and it will be more trouble than it is worth.

Chapter 8 - Finding and Securing Federal, State, County and Local Municipality Contracts

How to Find Government Contracts

This is where big businesses make all of their money, and so can you. The process for getting government contracts is slow but the payoff is typically very good. If you are new to this type of contracting then the first thing you need to know is how to find these jobs. The departments you are looking for can be found by searching for:

- Purchasing Department
- Procurement Department
- Doing Business With

The way to find which jobs or contracts are available is to use a search engine. We are going to use the city of Mobile, Alabama as an example.

To find what types of jobs are available in Mobile, Alabama you would search the city, state and the department. The term you will need to search is the "City of Mobile Alabama Purchasing Department". This will return the page that has all of the requirements and information that you need to start bidding on these city contracts and jobs.

Here is another example. In this case we are looking for the California Department of Transportation Procurement Department. In the search engine you would type "California Department of Transportation Procurement Department" and this will bring you to the page with all of the information that you will need.

Another way to search for these types of jobs is to know the exact website and search for those departments within the site. Here is a hint when searching. Google puts sites that are more established and have related content first. If you are looking for pages that have the exact title matches then Yahoo or Bing will work better. If you can't find the Purchasing department of a particular site then try this command:

inurl:cityofmobile.org "purchasing"

The inurl: command with the exact domain name tells the search engine that you are only searching inside that site and in this case it is *cityofmobile.org*. The quotations and the word "purchasing" tell the search engine that you only want pages with that exact word purchasing on them sent back to you.

Finally you can also just type "Doing Business with the City of Mobile Al". A lot of government websites have created "Doing Business With" pages in order to get more bids and get more competition in order to get the jobs done cheaper.

This system of searching can be applied to any city, state government or even business in order to find work. For example, General Electric has purchasing needs and so do most large corporations. If you type in a search engine "General Electric Purchasing Department" you will find everything you need to sell or contract for General Electric. If you want to learn more about different ways to conduct an online search, look up "Search Engine Operators". There you will find some pages that will show you all of the advanced features that are built into search engines to help you laser focus your searches. The smarter you work, the less you need to struggle to achieve the same results or more!

Another way of getting these types of contracts is to visit the SBA (Small Business Administration) or the GSA (General Service Administration). Both these site are government run and can direct you to where you can get the information that you need to bid on contracts. Be aware that the General Service Administration, or the GSA, is the federal government's Purchasing Department. That is all they do; purchase from businesses for the government.

Another place to contact is the United States Agency for International Development, or USAID, you can bid on jobs of all types, the products of which will wind up in locations all around the world. Go to **www.usaid.gov** to get all the information you will need to download bid documents or Requests for Proposals (RFPs), instructions on how to properly complete them, grant opportunities and so much more.

A perfect example of USAID at work would be the Ford heavy duty vehicles that were equipped in the USA by individual providers of items such as winches, dozer buckets, reinforced chassis, etc. These vehicles were for the specific purpose of aiding South American governments build their country's infrastructure. Those individual providers had won the bid from the American government to properly outfit those vehicles, which were eventually shipped out to Venezuela, Chile, Argentina, etc. by simply filling in the proper paperwork, and offering their most competitive pricing, equipment, and services. The government has already committed to providing these vehicles to a particular nation. Why not take full advantage of the existing opportunities and throw your hat into the ring? There is a lot of money to be made this way, but you have to go after it.

Additionally, the website **www.grants.gov** offers an incredibly wide selection of opportunities in all fields. There are a significantly large amount of federal government grant monies just waiting for the right organization or individual to come along, submit a proposal and win the job! On grants.gov you will also find all the information you need to learn how to go about submitting your proposals.

When learning how to find these types of jobs it is a trial and error type of research. Once you find all of the sites that are local to you then all you need to do is stay up-to-date on what is available. Most of the jobs available are not going to be for you or anything related to welding. As a business owner you need to go through the clutter and find the types of jobs that are right for your business to bid on.

There are also services out there that automatically notify you of contracts that are available in your industry and/or area. I do want to make clear that bidding on government and state contracts can be a full time job. If you can get to this point, it is worth hiring someone that specializes in proposal research and writing. Many mid-sized and large companies go as far as hiring private investigators to physically show up in their place simply because they don't want to risk losing a bidding opportunity. Some additional resources for you to look into include **www.USASpending.gov** and **www.FBO.gov**.

The 10 Rules of Government Contracting

Rule#1 - Paperwork or the RFP (Request for Proposal) must be in order and filled out exactly to the specifications outlined in the request or it will not be accepted properly.

Rule#2 – Be certain that you have the right insurance.

Rule#3 -Most jobs require a Performance Bond.

Rule#4 -Don't take on jobs that you cannot handle.

Rule#5 -Make friends in these organizations.

Rule#6 –Do not try to use urgency, high pressure sales techniques or anything like that.

Rule#7 – Don't try to bribe anyone, offer a referral fee or anything of the sort in exchange for special treatment. This kind of behavior could get you into some serious trouble.

Rule#8 – If you do not understand the contract or job, walk away.

Rule#9 – The lowest bidder always wins.

Rule#10– If you can afford a lobbyist, it is well worth the price.

The paper work *must* be in exact order. When bidding on these types of jobs it is referred to as a RFP or Request for Proposal. If it is not perfect, it will be thrown out without even being considered. Government, State, City and local municipalities require the paper work to be exactly right. This is the hardest part of bidding on these types of jobs! The people who review the bids receive insane amounts of paper work, and simply can't correct any mistakes. They are overwhelmed with work, and any bids that contain mistakes are quickly tossed into the recycling bin. Proper paper work is the golden rule when bidding on government jobs. You need to think about the poor people who are reviewing that paper work. They are overworked, under paid, and harassed by phone calls, the last thing they need is an incorrect application. Would you be willing to fix other peoples mistakes if you were doing that job? Probably not, right? Just complete the paper work correctly and give those poor people a bit of respect, because they deserve it.

Have the right insurance for a job *before* starting. In most cases you will need to show verifiable proof of the correct insurance before biding on or starting a job. Do not bid on a job if you cannot afford the right insurance. The last thing you want to do is get black listed before you even get a chance.

Many jobs require a performance bond in order to even bid on the job. First, you must find an insurance company or a bank willing to insure you. This type of bond protects the buyer from your potential mistakes, including your inability to finish a job or a bankruptcy. This potential risk is the reason new businesses pay a large premium. You need to shop around beforehand to determine if you can secure a performance bond prior to pursuing a big contract. In reality, bidding on any of these jobs would be a waste of time and money, if you cannot get a performance bond.

Do not bid on a job that you cannot handle! Just walk away and save everyone the trouble. You will be permanently blacklisted as a reliable service or product provider if you bid on and get a job, but then proceed to mess it up royally. Businesses that fail this way are well known within the business community, and there is no recovering from that type of failure. Do not make this fatal mistake.

'Making friends' is the king of all business games. This brings us full circle back to networking. If you are bidding on local or states jobs, go in person, and if possible meet the people reviewing the bids. Just drop in, and introduce yourself. Better yet, ask for some help by letting them know that you have never done this before. Do not ever pretend that you know it all, because in this situation, you will put yourself out of work faster than a horse during the industrial revolution.

Most government employees are fed up with contractors pretending to know everything and trying to tell them how they need to do their job. Honesty goes a long way, and saying that you have never done this type of work will likely buy you some forgiveness. Chances are someone will be willing to sit down with you, and give you some help or useful advice. Once they know your name and face there is an entirely different type of relationship established. Hopefully, you start a positive relationship that will go a long way towards your education and contact development in this arena.

You should already know the power of using an urgent situation to close a deal, if you have been running a small business for some time now. However, you should not use this high pressure tactic when pursuing government contracts. All that will do is annoy the people reviewing your bid and possibly get your paper work tossed into the trash can. They might say that if you needed that work so badly you should have spent your time pursuing the private sector, that pays right away. Please remember the people reviewing your bids make a small salary, are overworked and all they want is a break. Even if you could rush them, they have a boss that makes the ultimate decision! No matter what you do, it is completely out of their hands.

Bribing government officials is not a good idea. The point is you need to follow the rules, and yes, you can bribe people as long as it is a referral fee to someone working in an unregulated industry. The key word is ***unregulated*** industry. That would mean offering someone like a handyman a referral fee for sending a job your way.

On the other hand, you will likely spend time in prison or be liable for hefty fines if you should attempt to offer 'incentives' to politicians at the Federal, State, or City level. The only quasi-exception is if you go through a lobbyist that is paid to represent you and make a case as to why your company should get a contract. Lobbyists have their own rules and laws to abide by in order to perform their jobs, so leave it up to your hired gun to determine the best course of action to follow for each individual bid/proposal you are considering.

When in doubt, call your lawyer out! If you do not understand the contract, walk away, or pay your lawyer to explain it to you. Legal contracts are serious business and can land you in jail or court if you break, under deliver or overstep the legal bounds of your agreements. You need to know what you are signing and have protection against unforeseen events. The old saying, "A man who represents himself has a fool for an attorney" carries a lot of weight.

As a business owner, you need to know that the best way to protect yourself and your investment is to leave it to the professional you hire. You wouldn't want your attorney to attempt to complete some pipe welding work for you, would you? Why would you try to take on legal issues that you are neither qualified nor expected to know or understand?

All federal, state and local contracts typically go to the lowest bidder. Pricing is everything with government contracts, and the simple fact that the paper work is too much for most business owners to complete. This allows a big enough profit margin to justify bidding. A job well done, is a job well bid on. The bureaucracy of government contracts is what makes them profitable. Most businesses could not bid on them, unless the rules were not so strict. This is a huge advantage for you.

Think about hiring a lobbyist. Lobbyists are the ultimate investment in a sales force. They are typically lawyers and know what to do in all situations. They are the big wigs in getting contracts signed by government departments. Procuring your own 'hired gun' to go out there and fight to win as many bids as possible for you is definitely a worthwhile investment!. They argue your point of view with the ultimate goal of getting your business that contract. This is where all of the multinational businesses get their work. However, you must be prepared to pay a lot up front with no guarantee of success. This is the way to secure long-term, highly profitable government contracts.

Lobbying for Contracts – The Ultimate Return on Investment and The Big Payoff of Sales

Lobbying is a way to influence the Government to buy products and/or services from your business. This practice is dominated by large corporations that have insane profit margins because they have no real competition. Lobbyist go as far as writing and getting bills signed that will have your company in them with no price tag attached. You can charge whatever you want after you get the contract. No questions asked.

Lobbying is done by hiring a lobbying firm. What they do is go to the government officials and find a way to sell what you have to offer. Lobbyists can be just about anyone who can influence the government. In most cases, they are lawyers but it can also be friends and family of lawmakers earning a living because they can influence their friends or family for the right price.

For example, the wife of the governor of a state can lobby and influence her husband (the Governor) to hire a company to rebuild and repair a bridge. She would get paid to talk her husband into giving you that contract for the good of the state and the safety of its people.

Lobbying is done all of the time by corporations and special interest groups like General Electric, Northrop Grumman, Lockheed Martin, The National Rifle Association, health insurance companies, bankers, and human rights activists and so on. Lobbying is considered and protected under the freedom of speech and it is meant to allow specialty interest groups to be heard.

Lobbying has been proven to be the most effective marketing tool for big businesses. Here is an interesting fact. General Electric, or GE, spends over $100,000.00 a day, every day of the year, to lobby the Government for contracts. The eventual pay off comes when a contract is fulfilled and GE receives it's multi-million, sometimes multi-billion, dollar payment for a job well done. Here are some benefits of hiring a lobbyist:

- Government Officials Have No Concern for Cost
- The Tax Payers Are Paying the Bill and Do Not Know Where Their Money is Being Spent
- It is Estimated Lobbyists Get Their Clients a 22,000% Return on Their Money (that figure is NOT a typo! I triple checked it! 22,000% return on their money! No wonder so many companies are willing to pay these guys whatever it takes for them represent them!)

The main benefit of lobbying is that government officials are not personally responsible for the money they spend. In most cases, they do not care what it costs as long as the project makes them look good. If you were selling or biding on a contract to a private sector business, you would need to have the lowest price or the best time frame for project completion. In the private sector profit margins are small because the buyers shop around and watch every penny. On the other hand, government officials give out contracts simply because they either liked the project, have a personal or political interest in it, or they are returning a favor to a friend, family member, or business associate.

Having a contract paid by tax payers makes charging a lot more for the same work easy to do. No one is going to question the lawmaker's decision and the lawmakers will do whatever they can to not be made fools of. The tax payer is the one who will pay the bill and that makes it easy for government officials to sign on the line. No questions will be asked because it's not *their* money.

The return for hiring a lobbyist firm is high. It is estimated that lobbyists earn their clients a average of 22,000% on their investment. Let's compare this to other investments. You would be doing great if you can consistently earn 12% on your retirement investments. Most Investments pay less than 8% a year. If you market your business properly you can expect to earn $10 for every $2 spent on marketing and networking.

A lobbyist can average you a $220 return for every $1 that you spend with them. As a small business you can expect to spend $20,000 a year on marketing and networking expenses in order to gross $100,000 a year. Not a bad living by most standards today. But what if you were able to get the same return the big corporations get for their marketing and networking efforts? Let's say you could find a lobbyist to do the job for $20,000. You could expect to earn $4.4 million for the same year. That is a huge difference and that is why lobbying is the best return on your money if you can grow your business to that level.

Sub Contracting Government and Corporate Contracts

Did you know that if you miss or cannot get the larger government contracts, there is work available through sub contracting to the businesses that won those jobs? Here you get a kind of second chance to win those juicy government jobs!

The companies that win the big contracts must, by specific stipulations in their government contract, sub contract a percentage (about 25%) of the work to other small businesses. This is a built-in, 'pay it forward' system that helps 'spread the wealth' a little bit amongst America's small business owners. This is where you can win more work and have a huge advantage against the big businesses!

Don't be a shy! Find those contracts to get the job done. Just because you missed the main or primary contracts does not mean you are out of work. This is an even easier way to get work because the winners of the main contract have rules they need to follow in order to keep getting the big contracts coming in. In some cases, they are in desperate need to have some sub contracting done that they will even hold your hand and walk you through the proper steps just to keep the government happy.

To find these jobs you would use the same methods and similar search terms to the ones you used to locate government contracts. You first need to know what companies you are looking for. You can either find the companies that have won some big contracts or make a list of major corporations that are local to you. You can find this type of information in industry magazines, press releases, news articles, etc. Once you know the company name or website then you need to search for one or all of these three terms:

- Purchasing Department
- Procurement Department
- Doing Business With

Some examples of how you would search this is:

- Company name purchasing department
- Company name procurement department
- Doing business with company name

You can flip those search terms any way that you want as long as you specify the company name or website and what department or what you are looking for. Get online and experiment different ways of searching for these departments and once you find them make sure you either book mark them or save them to some sort of a list. It is exactly the same process as finding government contracts. Once you make a list then you need to keep up-to-date on what is available and what you need to do in order to bid for work.

Another idea is to follow the news as to who won the large contracts and then track down that company's purchasing department. You can either go through the government websites or you can search a news service for winners of these contracts. Here is a helpful hint: if you have a list of large corporations that are local to your area then you can set up a news alert from Google News to email you anytime one of these businesses has a big announcement. This way you have a head start. You might want to get your name on any mailing lists your clients may have available to the public. You would be amazed at how much valuable information many of your prospective clients give away that you can capitalize on if you pay close attention while getting to know what your clients are up to.

The Simple Sample of a 10-Step Marketing Plan

So now you have an idea of what goes into running a business. It is not easy, but you're going to make it happen because you aren't going to give up. Making it in any business is about dedication, following through on your plan, not being afraid of meeting new people and taking risks. As a business owner you need to understand that your job is to bring the business in the door. It does not matter if you do the physical work yourself, have employees working for you, or even sub contract the work out. It is your job to get work for the business.

If you don't want to do that heavy duty sales work that is required to get any new business off the ground and make it through those first few years, then you really shouldn't be a business owner. With all due respect, you are probably better suited and would be ultimately happier as an employee. An employee does their job without all the pressure and stress of being the boss, and all the worry that comes with risking your reputation, and perhaps even your life savings, on a dream.

Below is a general marketing plan and some suggested ways to spend your time and money. The key word is **suggested** because no business is exactly the same or has the same customers. First, you may have to change the way you interact with people and be ready to implement the techniques in this book, as they will put you ahead of your competition. You must also make a long-term commitment to become a better business person.

Before doing anything you must research all of the businesses in your area, particularly those that might need some welding work. Know your customer before you spend money.

Sample 10 Steps to Successful Marketing Plan

1. Create a company image with a logo, website, business cards, flyers, signs and uniforms.
2. Send out a press release about your grand opening with links to your website and put out the signs.
3. Make a list of all local businesses that might hire you.
4. Send out flyers to all the businesses on your list once a month for three months, and then every three months after that. Be consistent and soon your potential clients will all recognize who you are through sheer repetition if nothing else. Like any effective television commercial, the more often it plays, the more recognizable and memorable it becomes.
5. Advertise your business and website in industry specific magazines, papers, locations and websites. The key words are *"industry specific"*, which is usually fleshed out through a trial and error process.
6. Start visiting all of the businesses on your mailing lists with a business card, introduce yourself and say hello. Ask them if they need, or know anyone who might need, some welding or fabrication work done. The name of this game is contact. Sales is a contact sport and the more people you contact in person the greater your chances of success.
7. Find the people who can refer work to you and offer incentives or referral fees. Just make sure it is in an unregulated industry. You can approach other related sub contractors, project managers, handy

men, dock masters or anyone who has access to the type of people that may hire you.

8. Create incentives that will get businesses to hire your company. Once you are approached about your rates, quickly find out what the customer wants and give it to them. You are the deal maker and should do whatever it takes. Incentives are a great way to close the deal.

   Be prepared. Know what you can offer in advance so that you show no hesitation when making the incentive offer. They will appreciate your "quick thinking and decision-making skills, your obvious knowledge about their business, and your ability to listen to their needs". They don't have to know that it was all thought out beforehand!

9. Keep researching and updating your lists of businesses. Never stop contacting people and or advertising.

10. Once you have a list of established customers, stay in touch with them often so that they do not forget you. If you find that you can no longer fill this Business Partner Liaison position yourself as your business grows, make sure to hire a Liason officer in your stead. It is a lot harder to get a new customer in the door than to keep an old one!

Yes, it is that simple. Let people know about your business through the mail, then go out and meet them, so they know who you are, and finally close the deal. That is marketing and running a business 101.

The Marketing Experience You Already Have But Didn't Know About

Think about the last few times you've had to look for work. Isn't it all about deciding what kind of work you're after, knowing what your bottom line will be as far as salary and benefits, and marketing yourself to each potential employer? Need proof? Usually you send out your resume so they get a chance to know who you are, then you call or stop in for an interview and to find out more about the job, and finally you would close the deal with a weld test or a pay negotiation.

Want to know what the difference is between doing this as an employee verses being a business owner? As an employee you will do 80% of the work for 20% of the pay. As the business owner you will do 20% of the work for 80% of the pay.

You already have marketing experience in the welding industry. The big question is "who do you want to work for"?

CHAPTER 9 - WHAT KIND OF BUSINESS DO YOU PLAN TO BE IN?

Yes, you need a plan with specific goals and strategies that will help you reach them.

- Strategic planning takes place not only in a wood paneled boardroom, but in the garage-based, small welding business.

- You must always scan the business horizon for the right opportunities that will best fit your welding business.

- Dozens of viable business possibilities are presented in this chapter, you only have to choose one that fits your talents.

This chapter helps you focus your efforts and define your business plan. This is a very important part of your planning. You must think both long term, and short term about issues that face your company. The following pages offer you the opportunity to select a business that you feel is a perfect match for your abilities. Keeping in mind that whatever your choice, a business is a mix of products and services. You must provide both, and you must be driven to do quality work. The list begins with:

Independent Contracting and Staffing

Independent contracting or sole proprietorships are the simplest and easiest form of owning your own welding business. You can literally be in business and working in a few days with less than a $1,000 investment. As an independent contractor all you need are:

- Insurance
- Business Cards, Phone Number and Email
- Basic Equipment for the Job, however even that is not always necessary
- Independent Contractor Agreements and Invoices

This type of contracting allows the least amount of overhead and the ability to do the work cheaper than most established welding businesses. As an independent contractor you work just like a business except you give the customer an invoice for your work at the completion of the project, or at the end of the week or month. Taxes are paid by you at the end of the year and you use your social security number as your business identification number.

The main advantage that independent contractors have in comparison to other welding businesses is that the customer is able to employ a skilled welder without paying benefits, overtime or unemployment tax. As an independent contractor you sign an independent contractor agreement and you get paid on a flat hourly or by-the-job basis. You provide your own insurance and take some of the risk. In the end, you charge more, but do not receive the benefits or protection of being an employee. All of the terms of your work are stated in the independent contractor agreement. The down side of this type of a welding business is that if something happens, and you are at fault, then you are personally responsible. When the job ends there is no unemployment or benefits for you to collect. Some ideas for working as an independent contractor are:

- Staffing Manufacturing and Welding Shops
- Cruise Ship and Marine Repair Services
- Piece Work or Parts Repair
- Rig Welding

Doing staffing is an easy way to have your own business. This is no different than working for that business as an employee except that you charge more. A comparison of staffing business, as against being an employee is that an employee in a low end production job receives $15 an hour, plus overtime and benefits, while an independent contractor receives no benefits and charges a flat rate of $30 an hour and is paid 14 days after each weekly invoice is submitted.

A basic marketing plan for staffing local businesses is as follows:

- Create, and send out flyers to every business that does welding in your area on a monthly basis.

- Call or stop in to every welding job advertised and offer your services.

- Stop by every company that you sent out a flyer to and meet the owner, manager, human resource manager or personnel department head. Introduce yourself, give them your business card, and ask if they need your services. If you can afford to leave them a pen, or some other cool item with your company logo on it, even better! This person-to-person contact is most important. Sales is a contact sport! The more people you contact, the greater your chances of getting a job.

- Keep repeating these steps until you have more work than you can handle.

Cruise ship, yacht construction/repairs and general marine maintenance jobs are for independent contractors. They are not the types of businesses that market to the general public. For contractors, they are a networking business, run by relationships, and knowing the right people. Most of the welders and other trades working these industries are independent contractors, and come from all over the globe.

Cruise ship welding jobs in the United States are dominated by European and South American workers (English is rarely spoken). The main contractors will fly them in from all around the world in order to avoid paying insurance and other benefits like unemployment tax and overtime pay. As an independent contractor you will likely be working on the ship and that also means it is in international waters (even when docked at a US port), so US laws *do not* apply. To work as an independent contractor on a cruise ship you need to meet the companies that do the contracting. The cruise ship companies rarely hire welders directly and in most cases are not interested in American, Canadian or UK residents because of the higher labor cost. Most of the full time staff that lives and works on the cruise ships make less than $600 a month. What you need to do is go to the main ports and office parks to ask around and find out which contractors do the repairs and ship modifications. Just show up in person and hand out business cards. Once you meet the right people you will start getting calls for all sorts of independent contracting opportunities all around the world. As an independent contractor you typically get a lot of hours (100+ on a busy week) and in some cases only get paid after you return home. There are almost no living expenses on a ship because travel, room and board, meals and everything else is typically provided at no cost to you.

Yacht repair and construction is another opportunity for independent contractors. Most of the main contractors are not interested in full time employees, and only hire independent contractors to avoid paying overtime pay, benefits and unemployment taxes. In most cases, the tools and equipment are provided for you and you work just like any other employee except that you charge more per hour, and have no benefits.

The way to get into this type of contracting is by personally visiting every boat yard and marina, and handing out business cards. Start asking around and hand out lots of cards. The key is to make friends because this is a small international community steered by tight relationships. It takes a while to break in as a outsider, but after a while and enough persistence on your part, you will be hired. Even mega-yacht construction jobs are dominated by independent contractors because the investors don't want, or cannot afford, the overhead and insurance costs of employees. At this time, the hourly rate charged ranges between $20 an hour to $50 an hour, for an independent contractor. That is about double the rate of an employee doing the same job.

You can sometimes get work by the piece from a small business, if you have a garage with a welder. The typical jobs are either parts repair or small manufacturing shops that have a rare need for some welding. The key to finding these types of independent contracting opportunities is to find businesses that are small. Their size means that having a full time welder on staff would not be profitable.

RIG welding as an independent contractor is another form of owning a mobile welding business. As a rig welder you have your own mobile welder and charge by the hour. There are two rates charged. The first is for your time as a welder, and the second is for your equipment. These types of jobs are found by contacting other welding businesses and giving them your promotional materials and your business cards. You want to find businesses that have an occasional overflow of work and are not in a position to hire a full time employee, plus they need your equipment. Being an independent rig welding contractor is exactly like running a mobile welding business except you work more like an employee instead of a business owner.

Mobile Welding Services

Mobile welding services are what most welders want to do. Mobile welding businesses are a great way to start a business without too much over head (conserve money at all costs) or a great place to build a customer base before opening up a welding shop. Mobile welding businesses require the following to start:

- A Business Entity, Federal Tax ID or Business Number and Proper Licensing (depending on where you are located and the local laws)
- Insurance and possibly a Performance Bond
- Business cards, signs, flyers, advertising and a website
- Investment in a vehicle and portable welding and cutting equipment

Once you have everything up and going, your mission is to begin finding and contacting customers. As a mobile welding business you will most likely be doing general repairs. This means just about anyone who is involved with metal, is a potential customer. You can go two routes with a mobile welding business and they are:

1. General Welding Services and Metal Repairs
2. Specialty or Niche Welding and Repairs

General Welding Services and Metal Repairs

If you go the general welding route, you should market yourself to a variety of industries. The catch here is that you will have some competition. This is the area in which you will need to use incentives to get the work in the door. The main problem with running a general mobile welding business is almost always price or hourly rate. Most customers don't have a clue about what a good weld is as long as it holds. This is where many new welding business owners go wrong. They think that if they do excellent work they can charge the going rate (whatever it is in your area). You need to understand that the customer, with whom you will be dealing, only sees the hourly rate. You need to win them over by either charging less per hour or by creating an incentive to keep them as customers. Here are some examples of businesses and places where you can find general mobile welding work or jobs:

- General Contractors
- Structural, Civil and Mechanical Engineering Companies
- Crain Service Companies and Steel Erection Shops

- Architects and Designers
- Heavy Equipment Rental Companies and Services
- Mechanical Contractors
- Boat and Ship Yards
- Steel Mills and Metal Supplying Companies
- Sub Contracting Government Contracts
- Anyone involved in metals that may need welding services

The list is endless and you can find work just about anywhere. One of the things most business owners will say is you never end up working in places that you expected. The work or industry has a way of finding you instead. Here are some ideas for marketing and selling your services as a general mobile welding business:

- Create a company image with a logo and include that logo on EVERYTHING. Your business cards, flyers, signs, t-shirts, hats, pens, and especially your website. You want to create a brand that people will eventually recognize on sight.
- Make a list of every business in your area that might hire you. Start with the list just before this one beginning with general contractors.
- Put out a press release about your business and try to get a local paper to do a story about you.
- Mail every business on the list a flyer. Let people know who you are, what you do and that you want their business. If they do not know who you are,

what you do and that you want their business, how will they know to call you?

- Get out there and start knocking on these business owners doors. Stop in and introduce yourself, give them your business card, and tell them what you do. Just make friends, and ask if they need or know anyone who might need some welding services. A business owner that needs you will be more than happy to meet you. In most cases, these business owners are shopping for someone to do the job for less money than their current welder.

- If they are interested in hiring you because they have a job coming up, quickly find out what you need to do to get that job. If possible, avoid the business owner that wastes your time. People like this are always trouble. Stay away from them! A motivated customer wants the facts quickly, and is always ready to sign a contract as soon as possible.

- Keep sending out flyers at least once every three months and call or visit the businesses that you have met at least twice a year. If this feels weird to you then think of it this way. You are stopping in because you were either in the neighborhood, or are slow on work, and wanted to see how they are doing. Business owners love to talk about how their business is doing. Send out holiday cards or anything to let them know you have not forgotten about them can earn you lots of points. If you do not stay on top of your customer base someone new *will* step in and take your work away from you. Direct selling will enable you to meet and socialize with customers. Your earnings will be directly related to your efforts. It is your business, sell yourself, and the jobs will follow. People like doing business with people they like, and people they trust. Learn the customers birthday, send a card. Perhaps, the birth of a child, send a card. After a first meeting with the potential customer, a short note stating how much you enjoyed meeting them is definitely called for!

The whole point of a general mobile welding business is that you never know where the work will come from. Often it's a simple case of being at the right place at the right time. If your potential customers don't know about you or remember you, you *will* end up losing jobs to another business.

Specialty or Niche Mobile Welding and Repairs

Specialty or Niche Welding and Repairs businesses are run the same way as any other mobile welding business except for one thing. This time you are an **expert** and *the* person people want to do business with for your specialty or niche. This is a big advantage over a general mobile welding business because it means:

- Less Competition
- Higher Rates

- More Trust

As a niche welding business you need to create your marketing strategy and company image to suit the markets or industries that you plan to service. Chances are, you will cover a larger territory, but you can charge a much higher rate and there should be little to no competition. Here are some ideas and suggestions for a specialty mobile welding business:

- Restaurant Equipment Fabrication, Repair and Modification
- Hard Facing and Resurfacing
- Tool and Die Welding
- Chrome Heavy Wall Pipe Welding Services
- Welding Stainless Steel
- Welding and Fabricating Anodized Aluminum
- Welding Cast Iron or any Specialty Metals
- Cracked Boiler Repair
- Industrial Pump Station Crack Repair and Resurfacing
- Air Craft Parts Welding
- Marine Parts and Ship Repair
- Food Process Equipment Welding
- Crane Boom Repair and On-Site Heavy Equipment Welding
- Race Track Mobile Welding Services
- Gas Pipeline Welding

- Mirror Welding Services

As you can see in the above list, the more difficult or specialized the work, the less competition you will have. For example, if you specialize in mirror welding then you could easily charge two times the going rate because very few welders can do that type of work. This is an area that comes back to analyzing and knowing your market.

Bad ideas for specialty mobile welding businesses are:

- TIG Welding Experts
- MIG Weld Experts
- Stick Welding Experts

Bad ideas are any ideas that seem like a specialty to you as a welder, but the customer has no clue or idea what you are offering. For example, if TIG welding stainless steel is an area that you are an absolute expert in, you should promote yourself as an expert to industries that use TIG welding on stainless steel such as food process equipment, restaurant equipment repair or corrosive chemical pipe welds. The customer has no clue in most cases what TIG welding is and only knows they want an expert in the field they need repairs.

What you need to understand is that the person hiring you will most likely be an operations manager, project manager, office personnel, business owner or anyone who is in charge of writing a check to other businesses. What these people have in common is they typically know nothing about welding and will hire you based on their confidence in you. They like people who act, talk and look like they know what they are doing. If you actually know what you are doing, that is another story, and chances are if you do a bad job they will have no idea and only care that the work is done.

Here is a few examples of marketing and selling your services as a niche mobile welding business:

- Create a company image (or multiple images) that says Specialist with a unique logo, business cards, flyers, signs and a website. You want your business profile to scream "I'm a specialist!"
- Make a list of every business in your area that is involved in the industry you want to specialize in.
- Send out flyers or any other promotional material to let these businesses know about you.
- Put out a press release about your business and advertise in the publications that people in that industry subscribe to. If you are offering restaurant equipment repairs then advertise in a local restaurant wholesale equipment newspaper. Make sure you reach your customer.
- Go out and meet the people who might hire you. Remember people hire people and no matter how great your business is, the deal will be closed in person.
- Keep sending out promotional material at least twice a year. You are creating bonds in order to have a

long-term relationship with the customer. Sell the benefits of your service. Make certain that you provide the very best customer service. Talk to the customer, learn their unique needs and meet those needs. Constantly ask yourself, do I really know my customer?

As a niche or specialty business you have many advantages. The main advantage is the trust people have in your expertise. It does not matter if you know anything about the business or that all weld repairs are almost the same. It really comes down to the customer or buyer having confidence in you through the marketing that you have done.

Welding and Fabrication Shop

Before even considering opening a welding shop you need to understand the risk and know the alternatives. The risks associated with opening a shop are:

- Committing to lease
- Paying for insurance, permits and licenses
- The cost of equipment
- Running out of money before getting established

Opening up a shop means you will likely need to rent shop space. Renting or leasing commercial space is a tricky business and is nothing like renting an apartment or buying a home. This is an area that you need a lawyer, or at bare minimum, a commercial Realtor who specializes in commercial leases. If you make a mistake here you will be paying for many years to come. Remember leases are legal binding contracts and whatever the lease says is what a Judge will rule in court. There are no excuses like "I did not understand the lease" or "The owner promised me this". As a business owner you are now considered a sophisticated investor, and the laws no longer protect you. Think things through, have an expert look at it for you, and try to negotiate some concessions *before* committing to any lease.

Any lease that you sign will require having the proper insurance. Besides the cost of insurance there are going to be other fees from local and state agencies. For example, an occupational license, and a fire safety inspection from your local fire Marshall. Some cities like New York City require welding and cutting equipment to be accompanied by a formal permit and an operator licensed by the city to operate it, while in places like Montana there exists almost no red tape or safety requirements. The bottom line on insurance, permits and licenses is to have a lawyer or accountant advise you as to who you need to contact before opening up your shop.

Opening a shop will require buying, renting or leasing metal fabrication equipment. Ideally, the best option would be for you to rent the equipment on an 'as needed basis' only when called for. Your next best option would be to buy used equipment for much less than you can buy new. You can always upgrade later. If you commit to new or leased equipment, you've just committed money that could have been spent more wisely elsewhere. Just a reminder that equipment sales are a place where predators make a killing selling to new business owners at high prices, and then buy that same equipment back in a year, for pennies on the dollar.

Opening a shop and going for all of the bells and whistles is a dangerous game because it requires committing your money to something that does not give anything back for a while. As a new business owner you need to save as much cash as possible until you start getting enough work to justify spending. If you spend too much in the beginning you will put yourself out of business and end up getting a job to pay for those losses and have absolutely nothing left to show for your efforts.

Alternatives to Initially Opening a Welding Shop

A less risky alternative to opening a shop from day one would be starting a mobile welding business to build a reliable customer base. Later on you can make a deal with an established welding shop to either rent a small space in their shop, use their equipment, or even partner deals to get access to their employees. This allows you the luxury of having very little overhead while at the same time allowing you to take on even larger projects than you would have been able to handle if you had opened up a small shop from day one. If you really want to be a business person, you could simply sign the contract for the work and subcontract it to a larger welding shop that is struggling to pay their bills. As a new business owner you need to find creative ways to keep costs down in the beginning until you can justify spending.

General Welding Shop vs. Niche Welding Shop

A general welding shop is open to just about any type of welding and fabrication job you can handle. It is a competitive market, but over time you get customers from just about every business sector that you can imagine. It is a safe business to run and only requires an owner who can get out there and bring the work in.

A niche welding shop is the type of business that you market to specific industries. You are an expert in the field you choose and people trust you for that expertise. You can charge a lot more than a general welding shop and your customers are basically forced to send you jobs from all around the world. A niche welding shop can take longer to establish, but once it has a customer base, it is a very solid and safe business. Why? Because most of your customers have no other, 'highly skilled' alternative to chose from. When you are the only game in town, sooner or later the people who need you will find you.

Marketing plans for welding shops are the same as a

mobile welding business. You need to create a business image, market your business and then reach out and meet your potential customers. All businesses are about relationships. Remember this; putting up a sign will not get you anywhere, only you can get business in the door.

Metal Manufacturing Business

Starting a manufacturing business is an excellent idea because you control the work, type of product and don't rely on a customer to strike an arc. You can experiment, build what you want, create any product or even works of art that you feel might improve people's lives or that you feel will simply become best sellers. The options are endless and there is always demand for a better product.

As a manufacturing business you can keep your overhead and commitment down by registering your business and using a P.O. Box as a business address. All of the manufacturing can be done from a garage and products can be stored in a storage unit. It can be a start of a very profitable career and you never know how far you can go. You must carefully think about, and research what can you manufacture to sell. That is the hard part that only you can answer. You can create a better product, manufacture one that is cheaper or buy the rights to manufacture a product as the new owner. Those are you three choices.

Another great thing about having a product to manufacture when you have a welding shop is that you are able to pay the rent on a regular basis once you have customers. This is a huge relief for most business owners and allows you to work on getting more business in the door, expanding your product or creating newer and better ones.

How to Sell Your Product

So you have made a product and strike an arc anytime you feel the urge because you own that product. The million dollar question is "How do I sell it?" That is easy! You have two choices and they are to sell:

- Directly to the Customer
- Through a Wholesaler, Distributor or Store

The most profitable way to sell your product is person-to-person selling. You would utilize marketing and a sales staff, just like any car dealership or infomercial. Another option is attending trade shows to display your products and take orders. On the other hand, if you happen to have a product that is very useful and could be in high demand, you could lease the rights to a larger manufacture to produce and market it on your behalf. Be warned that this tack can be very difficult to accomplish, could be considered to be a huge gamble, but if all goes well, it could also be the way to make your fortune!

Selling through wholesalers, distributors and stores is

not as difficult as it seems. The people who sell products know you are not a marketing expert and need you as much as you need them. There are two ways to go about selling your product:

- Directly to a wholesaler, distributor or store
- On consignment

Selling Products Directly to Wholesalers, Distributors and Stores

Dealing directly with wholesalers, distributors and stores is not difficult. Finding the buyers of products comes down to knowing your market and contacting the right magazines, shops or businesses. The department that you need to contact is the "purchasing department" just like getting a government job, but without the red tape. They either ask for your product as a sample or say no. It's that simple. The market for selling your product is endless and all you need to do is find the right people who want it. The only down side is you don't get paid for a while and that all depends on the sellers rules. Some businesses pay you in as little as 30 days while others may want 6 months to pay you for what you gave them today. Don't be shocked because this is how the entire business world works. Therefore, you must be prepared to make a sizeable initial investment in order to 'front' the products that are to be sold.

One more available avenue would be to hire a <u>recruiter</u> who enlists other distributors to work for you. When a distributor sells an item, you receive a previously agreed upon percentage for each item sold.

Selling Products on Consignment

If you build a product and want to sell it locally, or even nationally, you have the option of offering it on consignment. All this means is that you would find a local or national seller and say, *"I have this product and I want to sell it, I want a certain amount of money from each sale, and whatever amount you're able to get above that you get to keep."* It's that simple. You give that business your product and if it is sold, you get your money and if not you get the product back. If they sell it, you give them another and so on. It gives businesses a product to show while saving them the expense. It is a win-win situation.

General Welding, Fabrication, Machining Business Ideas, Industries and Suggestions

Opening up a general welding, fabrication or machining business have one thing in common with a general mobile welding business. They both have a similar customer base. This customer base might provide your income. Here is a list of suggested business types to approach for work. Although it is a generic list, its purpose is to provide ideas and inspiration. You never know where your work will come from.

- Architects and Home Builders – These businesses always need metal stairs, railings, sheet metal work, and all sorts of metal products.

- Engineering and Mechanical Contractors – These both need pipe, structural products and custom metal shapes.

- Marinas and Ship Yards – These two always have work for two reasons. Primarily, they have no skilled welding or fabrication staff, and secondly they are typically short on people when big jobs come in.

- Government Contracts or Sub Contracting – There is no way to know what you will find until you start looking.

- Transportation and Heavy Equipment – These industries always need welding and custom made parts.

CHAPTER 10 - NICHE WELDING BUSINESSES, MANUFACTURING SHOP IDEAS, INDUSTRIES AND SUGGESTIONS

What follows is a general list of ideas to assist you in thinking about the many different ways you can make it in a welding business. The goal is to expand your view of what a welding business can be, and to provide you with some more focused ideas of where you may want to start your business. You don't really want to be doing the same things everyone else is doing because that will only lead to working for less money than your competitor. Try to remember that the less competition you have, the greater chances of your success because you can charge more for less work.

Think back to welding school where they said that a career in welding can take you just about anywhere. Welding is involved in just about every industry, and you should find your place within it. Niches have a funny way of finding the right people for the job, and you need to accept that. For example, I never thought I would write a book about welding or even cared about writing (I actually hated writing). Five years ago I could not write a two sentence add for the classifieds and now I wrote this book. I am still amazed at how I got here because you never know where your career in welding will end up. Take your time, be open minded, don't assume anything and let your welding business tell you where it needs to go. Listen to your business and customers! If your business keeps getting request for stainless steel work then you may want to take that as a sign that your customers seem to think you are the right person for that type of work. Just listen to the market and your customers.

Being a niche welding business is just like being the person on the jobsite who has a special skill. If you are the person who can weld a open root pipe with a really bad fit up that is 2 inches from the wall and a inch from the floor while using a mirror and being able to pass x-ray, then you are the undisputed expert in tough situation like that and people will always turn to you for that expertise. Don't forget to charge a premium for your skills. You can do as you please on the job, and no one will question your rates, or have unreasonable requests, because you have the answer to their problem. Here is the list with some basic ideas on how to get involved in niche welding businesses:

Aircraft Parts Repairs, Fabrication, Welding and Remanufacturing

Aircraft have expensive parts and there is always a need for repairs. As an expert in aircraft welding you will need to be certified to the codes that are needed and you will need to know where you can find work. The up side is the work is typically clean and pleasant. Some places to find work are:

- Small Airports and Hangers
- Aircraft Service Centers
- Aviation Companies
- Aircraft Auctions

When finding these places it is always best to look for businesses that don't have enough work to justify hiring a full time welder. You can find jobs by meeting the people at your local airports, hangers and aircraft specialty shops for repair and fabrication jobs. Some welding business owners buy broken aircraft parts at auctions for pennies and repair them. The catch with buying and remanufacturing parts is that there is typically a license/certification that is needed in order to resell the part. After being certified you can sell repaired or remanufactured parts online or to dealers. It can be a good business for a small shop or one person operation.

Aluminum Fabrication, Repairs and Welding

Aluminum is one of the easiest metals to work with. It is light, easy to shape, bend and machine while maintaining a low rate of distortion and it happens to weld great. However, few shops specialize in it. Specializing in aluminum seems like a big deal but the work is easy and customers always trust experts in that material. Aluminum requires very little specialty equipment and takes minimal training to weld. As an expert in aluminum you get a broad range of customers without having an exact niche. It's like being a general welding business with a niche but a wide range of industries that you service. Some places where you can find work or sell products are:

- Marinas and Shipyards
- Boat Dealerships New and Used
- Architects
- General Contractors
- Truck Stops
- Heavy Equipment Dealers
- Chemical Plants and Manufactures

As an aluminum welding business you can hit just about any industry that deals with aluminum. There are lots of options as to where you can build a customer base and expand. Primarily, you need to be known as the expert in that field. Send out flyers and meet lots of people who are potential customers. Other options for working with aluminum are products that you can manufacture under your own brand, patent or trademark. This means you can create a steady income that will grow over the years while still working a welding business. There are many well known product brands that have started as a small business and now all they do is manufacture their own equipment.

Architectural Iron, Gates, Handrails, Stairs and General Safety Products

Architectural iron, gates, handrails and stairs are needed in just about every part of the world. It is the type of niche that is needed by everyone from homeowners to industrial complexes and even ships or Government projects. This is one of those niches you need to thank lawyers for all of the slip and fall lawsuits and insurance requirements. In this niche you can either do the work as a client requests, or you can start your own line of products. There are lots of opportunities here and even though there are many other welding businesses competing for these jobs there is more than enough work to go around. Some of the places you can find work for this niche are:

- Architects
- General Contractors
- Town, City, State and Government Contract Jobs
- Yacht Brokers
- Ship Brokers
- Marinas
- Home Shows
- Medical Supply Stores and Hospitals
- Safety Supplier Stores and Catalogs
- Industrial Construction Contractors
- Airports and Aviation Terminals

Most architects, home builders, general contractors,

marinas and government jobs need all sorts of gates, hand rails, stairs and in some cases artistic architectural iron work for their projects. This is a niche where you can find lots of jobs in urban areas by working with the artistic crowds. You can find work in places such as Home Depot, Lowes, or by going to the home shows and showing off your work or even making standard products that are either sold in stores or bought directly by contractors. If you have that special touch and are artistic there are very high price tags for original welding art work done for specific projects. The opportunities are all over the place for the right business. It just takes some creative thinking as to where you can find the work in your area.

Automotive Parts Repair, Fabrication and Manufacturing

Almost every city, town or county has lots of automotive repair shops. All of these have one thing in common; customers that need a part but can't afford a new one. By being an expert in automotive parts welding and repairs you can get lots of work and don't need to market your business for too long. You can provide some much needed service to customers that will be more than happy to welcome you. The down side is welding automotive parts can be tricky. You will need a good parts cleaner to get all of the dirt washed off, and in most cases you will need to preheat or at least know how to properly fix cracks and proper metal cool down procedure. It is a niche business that has lots of demand for the right business owner. Some areas where you can look for work are:

- General Automotive Repair Shops
- Automotive Machine Shops
- Custom Automotive Accessories
- Speed Shops
- Race Tracks
- Used Tire Shops

As a welding shop owner there are many areas to choose from. You will find that there are lots of opportunities from business owners who are not being taken care of, if you choose the automotive welding field,. The welds are dirty but people need them done. Think about this. An engine shop throws out ten engine heads on a weekly basis because they are cracked. These heads are worth at least $200 each in repair cost, if they can be fixed. If you had the right skills you could either charge the client for a repair or take the cracked head, fix it and resell it as a remanufactured head.

Tire Shops would be yet another area where you could cash in. There are many tire shops that throw out cracked aluminum rims. You could offer to repair them for less than a quarter of the cost of a new one, and the shop owner can still charge a fee on top of that. If you do this full time, eventually you'll be able to invest in equipment that would allow you to completely remanufacture cracked aluminum rims. You can take the business as far as buying cracked rims for pennies on the dollar, and remanufacturing them so they can be resold at almost full price.

Speed shops and race tracks always have people that want custom exhausts, parts, roll cages and racing accessories. You have the option of just fixing stuff, buying and repairing for resale, or starting your own line of automotive accessories. The opportunities are only limited by your imagination.

Air Conditioning, Heating, and Plumber Contractor Welding Services and Fabrication

This is a niche that is virtually ignored in most parts of the country! A/C contractors need welding services but you need to find the right ones. There many A/C contractors who have welding work but refuse the jobs because they have no welders. Commercial A/C units need chill water pipes welded (6010 root/7018 hot pass and cap) as well as commercial custom A/C condenser stands. The catch is obvious here because you need to find the contractors who have just enough clients that they don't need to hire a full time welder. If you are considering getting into the fabrication/manufacturing area, then you can make a series of customizable stands and accessories for A/C units! The work is there but you need to find the right people to hire your company.

Heating contractors need custom duct work and general welding services. Small heating contractors typically turn down jobs that require welding and if you make the right connections you can partner or sub contract the jobs that require welding. You can expect to work on fuel tanks, boilers and pipe systems.

Plumbing companies and mechanical contractors occasionally need pipes and fittings to be fabricated and welded. The ideal business to contact does mechanical and plumbing work but does not offer pipe fitting and welding services. These are small jobs, but if you get enough business, then you can make a nice living. Besides that these businesses sometimes need general welding work done like handrails, safety platforms and stairs.

ATV, Motocross, Snowmobile and Motorcycle Accessories, Repairs and Trailers

ATV's, motocross bikes, motorcycles, snowmobiles, and the like have many needs from the welding community. They are all customized in many ways as an expression of the owner's personality and everyone wants a unique ride. Here are some examples of what you can do:

- Custom Frames
- Performance Parts
- Bike Modifications
- Repairs

- Custom Trailers

Admittedly, this is an area that does not have much work available unless you offer a unique service. It comes down to either manufacturing your own product line, getting in with shops to do their custom work or making a name for yourself as a specialty service. As an example of a specialty service, you could stretch-out ATV's or take stock street bikes and turn them into custom choppers according to the customer's order. The possibilities are endless. One of the more common areas is building trailers. There is not a lot of money in it, because it is production welding. On the other hand, building trailers can be profitable if you set-up your shop right with jigs and create a factory floor. Then it would be a profitable business that has the same customers giving you orders every month. The places where you can find these customers are:

- Motor Sports Dealerships
- Off Road Vehicle Rental Companies
- Motorcycle Dealerships
- Race Tracks
- Accessory Supply Catalogs
- Online Forums
- Riding Groups
- Magazines that are devoted to the riding/racing enthusiasts

There are opportunities here. All you need to do is find your place and create something people want. Besides welding, you also get to play with a lot of fun toys and you can buy bikes that become a tax write off for your business.

Automated Welding Equipment and Robotics

Building robotic welding equipment is an area that will require a bit more than just welding skills. But for those who like to design and tinker, this is a very profitable business, and a challenge that combines welding, electronics and coding. Many businesses cannot stand to pay an employee for doing their job, but they are more than willing to spend insane amounts of money to eliminate a job. Think back to the $20 ball point pen. You just need to be there to collect that money.

Designing automated welding equipment also has many grants from the government that you can collect to get your business started. Since welding is a huge part of the defense industry there are always going to be funds to improve the quality and speed of the welds the defense industry needs.

The buyers of these services and equipment are typically large manufacturing businesses or industrial construction contractors. You will have more work and profits than you will know what to do with, if you can design and manufacture robotic welding equipment that can streamline or completely eliminate many welding jobs off the factory floor. At worst, you can receive some grants to fund your research in designing better welding robots.

Bathroom Fixtures, Accessories and Custom Handicap Safety Equipment

You might not think much of fabrication and manufacturing bathroom accessories, but there is a demand for custom fixtures, handicap safety equipment and just about anything metal for the bathroom. This is an area that few people are servicing and there is the potential to sell your products on a national, even international, scale. Some places where you can find buyers and work are:

- Architects
- Interior Decorators
- Kitchen and Bathroom Contractors
- General Contractors
- Hardware Stores
- Home Improvement Stores
- Medical Supply Stores
- Supply Catalogs
- Home Improvement Magazines
- Industrial Suppliers
- Hotel and Motel Chains

All of the above need good stainless steel and chrome fixtures and have very little choice in the market place. If, for example, a contractor who is involved in the construction or renovation of a one hundred room hotel might require a unique look for their bathrooms. The contractor is going to need custom shower curtain holders, towel racks and handicap safety equipment. The best part is that much of this equipment is required by law and any commercial building will require it.

Perhaps specializing in stainless steel bathroom safety equipment for the handicapped is another area you may want to consider for yourself. You have a product and service that can literally change the lives of people with special needs.

You can build products that you can sell to stores like the Home Depot, Lowes, Ace and local hardware stores, if you want to do is strictly manufacture. All it takes is the right product and getting in touch with the right person in the purchasing department. This is a great niche because there are lots of work options and it can be done by a one person business.

Boating Parts, Accessories and Repairs

Boats are nothing more than a hole in the water that owners love to throw money into. That is exactly why it is a great industry to service. Boat owners love their boats and spend lots of money on. Some ideas of what you can offer to the boating industry are:

- Propeller Repair
- Custom Ladders
- Tee Tops
- Tuna Towers

- Marlin Towers
- Boat Lifts
- Dock Accessories
- Fishing Accessories
- Stands and Supports
- Trailers
- Swim Platforms and Floating Docks

There are a multitude of opportunities in the boating industry which is mainly serviced by small businesses. Larger companies typically stay away from the marine business because so many of the products are handmade. This means that no two of the same products are exactly the same and therefore cannot be completed via production line. The upside is that the products are mostly custom made and there are no standards in the boating industry. An example of this would be a yacht. Even though they are made in a factory, nothing is square and every side of the yacht is a little bit different. That makes mass producing and cost estimating add-on products very difficult. As a small business, this is the ideal situation to work in. Some places that you can find work are:

- Marinas
- Boat Dealerships
- Marine Boat Catalogs
- Boating Websites
- Boat Shows
- Magazines

- Boat Clubs

There are a few options to getting involved in the boating industry and they are:

- Repair Work
- Custom Fabrication
- Manufacturing a Product Line

Repair work is pretty common but it all depends on the size of the boats in your area. If you live in a place where boats are small, then the repair work will be scarce. On the other hand if you live where there are many waterways and ports then there will be a lot of welding repair work needed. It goes back to learning your market area and knowing who needs your services.

Custom fabrication is always needed in the boating industry. The customer base ranges from the boat owner to the local dealerships. People need tee tops, towers, bow rails, fishing accessories and all sorts of stuff made to fit their boat. This is a very profitable business and more than enough work to go around.

Finally, you can start manufacturing your own products. There are lots of boating and boating accessories that you can manufacture and sell. The list is long and the best part is you can sell locally or even go global. If you really want to make a great impression you might be one of the few businesses that offer 100% anodized aluminum products. Most small businesses are too cheap to buy an anodizing tank but many dealerships would love a product like that. Besides that, the money saved on buying anodized aluminum would be worth it and the welds no longer need to be painted. Your buyers will find you through your selling directly through magazine ads, selling to stores and dealerships, attending boat shows, or even giving your products to dealers and shops on consignment. If you live in the right place, you have options.

Boiler Construction, Repair, Modification and Welding

Boilers are big business and always in demand. It does not matter if you live in a warm or cold climate. There are boilers all around. In cold climate areas most buildings have boilers and a need exists for repairs or replacements. In the warmer climates there are always factories that have boilers and need them working continuously. If you specialize in boiler construction, repairs or modifications they you can have a very good business. The catch is you will likely need an ASME 6G restricted welding certification and a really good insurance/performance bond policy. It is a high risk area and that is why it pays well. Some places you can find these types of jobs are:

- Plant Superintendents/Engineers
- Building Superintendents
- General Contractors
- Government, State and Local Contracts

- HVAC Contractors
- Procurement and Engineering Contractors
- Plumbing Contractors
- Mechanical Engineers and Contractors
- Staffing Companies
- Property Management Companies or Community Association Managers (CAM)

There are lots of job opportunities in this niche, however you must be able to travel for work. The only exception to this is if you live in a larger city but then it is run by union contractors and you will have very limited options. Another option is manufacturing your own boilers or fuel tanks. There is some red tape to get through, but a lot less than you think. Some areas you can specialize in are:

- Boiler Tube Welding
- Mirror Welding
- Custom Fuel Tanks and Stands
- Boiler Crack Repair

All you need to do is let the right business owners or employees know what you can do for them if you have the skills in any of these areas.

Bunker Building

Bunker building is an area that few businesses cater too. There is a growing trend of people called 'preppers' out there who are preparing for a major disaster to occur, and many of them want steel bunkers made in which their families can bug out to and ride the storm out as it were . The price tags for these jobs are pretty high and start at about $20,000. Building bunkers is a niche that has a strong loyal following that will do just about anything to get the product they want. These prepper communities are unique and have all sorts of materials (magazines, blogs, websites, tv shows, etc.) that you can use to advertise your products and services. Be aware that one of the qualities these people are seeking above all else is your ability to keep the work you do for them (especially bunker locations) totally secret.

Commissioned Metal Art

Commissioned art sells for big money and is in demand if you find the right buyers or crowds. Other terms for this type of art are:

- Corporate Art
- Industrial Art

These jobs are found by either finding the right places to meet people who are buyers of this type of work or looking for corporate or city projects and offering your services. The good thing about this type of work is that it is 100% unique and you are free to create whatever you feel is the right style for that job. If you happen to create some art work that is outstanding, and you develop a following then, you will be making more money then you will know how to spend.

Cages for Animals

How many welding businesses do you see advertising cages for animals? This is another area like boating where the buyer expects to lose money but does not care simply because is a personal passion. This is a niche where you can work from a small shop and ship your products nationwide and pretty much make your own schedule. You can get these types of jobs advertising in magazines and papers for pet owners. If you want you can create your own product line or build to suit on an individual basis. It is an easy, one person niche and there is demand for these types of products. Just think, if you bought an exotic cat like a Tiger, then where are you going to find a cage for it? If you can afford exotic pets then a few thousand on a cage is not a big deal.

Chemical Plant Welding and Fabrication Services

Chemical Plants are always in need of piping, processing equipment repairs, modification and fabrication. This is an ideal niche for mobile welding businesses that are willing to travel. Most of the work does go to either large contracting companies who have deep pockets, or the work is done in-house. The ideal situation would be to offer specialty welding services that deal with metals that do not corrode, steam pipe and vessel welding or even boiler tube welding. This is a general welding business niche but when you add the words "Chemical Plant Welding and Fabrication Specialists" then the people writing the checks have more confidence in you.

Another area within this niche is to simply offer specialty welding services and nothing more. In many cases of plant shutdowns, the contractors can't hire enough welders and they will do whatever it takes to get the job done. Plants that shut down can lose millions of dollars a day when they are not in operation. Their time frames and deadlines are set in stone and this becomes a "do-whatever-it-takes" situation. Some of the areas that chemical plants have a shortage of welders in are:

- Welding Chrome Heavy Wall Pipe
- Thin Wall Stainless Steel Piping
- Boiler Tube Welding
- Mirror Welders

Any of these areas are always in demand as long as you are willing to travel. In many cases a contractor will keep you on staff and pay you for your services and your rig simply because they can't afford to lose you to another contractor. This work is no different than being a road warrior except you can charge a lot more for your time.

On the fabrication end of this niche you can find work by subcontracting from the general contractor or engineering company. You need to find the businesses awarded the contracts and then go through the proper bidding procedures to win the work.

Dairy Equipment Repair, Manufacturing and Welding

Dairy equipment is always in need of upgrades and repairs. It is almost exclusively made of stainless steel and must be sanitary. This is a niche that is great in areas with lots of dairy farms or to those who can manufacture useful products to make the industry cleaner or more efficient. The dairy industry is always in need of:

- General Repairs
- Process Piping
- Holding Tank Fabrication and Construction
- Specialty Fabrication Items

General repairs are very common and they range from cracks to replacing small portions of equipment. Most dairy equipment is welded so this industry relies heavily on TIG welding stainless steel. Besides that, everything must be polished so no areas can build up bacteria.

Most dairy plants and farms need to replace their pipes every so many years and this means lots of repeat business. It is typically Schedule 5 or 10 thin wall stainless that is welded but there is a lot of it. There are two ways to get this type of work and they are either to bid on the jobs or subcontract from a larger engineering company that has a work overload.

Another area is the manufacture and fabrication of storage tanks. These can be tanks that are put in farms, dairy factories or the tankers for the trailers that trucks deliver the products with. Either way, there is lots of welding work available plus the price tags for this type of work can be quite lucrative. You can find this type of work by contacting the following types of businesses:

- Dairy Farms
- Dairy Product Producers
- Transportation Companies that Specialize in Dairy Product
- Ice Cream and Cheese Manufactures
- Food Processing Engineers and Contractors

Exercise Equipment Fabrication and Repair

If you live in or near a large city, you may want to advertise a part of your business to gyms and gym equipment wholesalers. Gyms are always opening and they sometimes need custom racks for their weights and general weld repairs. There are also many people opening mobile personal training businesses. They typically buy a large box truck, load it up with gym equipment and go to the client's house. These trucks need lots of custom metal work to keep the equipment from moving around when driving.

Another option is specializing in custom weight racks and builds to suit equipment. For most people in the gym business it is a lifelong passion and they will go to extremes to get what they want built.

Farm Equipment Repair and Fabrication

Farmers always need repair work done, and in some cases custom fabrication. This is the way to go if you live in a farming community. You can offer general welding repairs, custom fabrication and build custom farming equipment if you have the necessary expertise. Farms come in all types and there are lots of opportunities. Some examples of services you can offer include:

- Dairy equipment repairs, modification and fabrication
- Storage tank fabrication
- Tractor and heavy equipment repairs and hard facing

General farm welding service work can be found just about everywhere. The trick here is to do more specialty work such as working on dairy equipment, building storage tanks or hard facing, high abrasion equipment.

Fuel Tanks

There is a lot of demand for fuel tanks and many businesses need custom, made-to-order fuel tanks. Some welding businesses do nothing more than build fuel tanks for vehicles and boats. Heavy equipment operators need them to transport diesel to their equipment. Homes need them for their boilers. Boats need them to fit in odd spaces. Many people need custom made fuel tanks for many reasons. The bottom line is that there is a ton of work for your shop if you want to get into fabricating fuel tanks. You could even go as far as making a product line.

Furniture and Custom Décor

Furniture and custom décor is an easy, one person business that can pay very well. There are two ways to go about doing it and they are:

- Start your own product line.
- Offer custom work.

If you start your own product line then you are free to work and do as you please. You will need to find high-end retailers that are willing to sell your products. You might just be able to charge whatever you want if you are able to create trend-setting styles. Think of it this way; if designer shoes can cost well over a $1,000 a pair, what do you think a high end piece of furniture that's in demand can sell for?

Offering custom furniture and décor can also be a very profitable business. The one thing wealthy people love to brag about is owning a one-of-a-kind piece. Think of this the same way you would in terms of a custom hot rod or chopper. Get the picture? You can find these types of opportunities from the following:

- Architects
- Interior Designers
- High End Furniture Stores
- Home Decorators
- Hotels
- Restaurants
- Country Clubs

Gardening and Home Improvement Accessory Ad Ons

If you are able to design accessories for gardening equipment, or add-on's, then there is another national market you can tap into for profits. Some examples are small plows for small tractors, mower attachments, jigs and saw mill guides for chain saws, and anything else you can think of that is original and sought after. I recently found a business for sale that makes aftermarket products for John Deere Gator Products. It earns $230,000 a year net for the owner and that's with three employees on the pay roll. All you need is the right idea and a place to sell your product. It is not an easy market to break into, but it is one more business possibility to think about.

Hard Facing and Surface Building

This is an area that has lots of work and it does not matter where you live. Businesses hate spending money. They would rather rebuild than to buy it new again. Besides that, hard facing lasts longer than most steel products. Some areas where you can find this type of work is:

- Stone Quarries
- Heavy Equipment Businesses
- Miners
- Dump Truck Owners
- Heavy Equipment Rental Services
- Machine Shops
- Factories
- Aircraft Machine Shops
- Recycling Plants
- Government, State, City, and Local Municipalities
- General Contractors
- Shipyards, Marinas and Boat Yards

Hard facing is a specialty welding service and there are lots of opportunities from many industries. Any type of equipment that has contact with earth wears down quickly from the friction and it needs to be rebuilt in order to keep it running properly. Actually, any industry in which the working metals are worn down regularly and need to be re-machined would be a good target client for you. It costs a lot less to build up a surface than it does to buy a new piece of metal and start from scratch. For example, jet engines have their turbine blades rebuilt every so many hours; industrial pumps are resurfaced and re-machined every so many years, and equipment like excavators need their buckets and tracks resurfaced to keep operating. This type of welding service saves business owners a lot of money and it is all about repeat customers.

Hydraulics Parts Fabrication, Welding and Repairs

Hydraulic parts fabrication and welding is a business that has a decent amount of demand just about anywhere. There is demand from hydraulic welding and fabrication services from:

- Hydraulic Hose Manufactures
- Heavy Equipment Operators and Contractors
- Marinas, Boat Yards and Ship yards
- Factories
- Crain Services
- Mines and Quarries
- Specialty Automotive Shops
- Heavy Equipment Remanufactures

Most of the work in this area is welding hydraulic connectors to pipes so that either someone can resell them or use them for a replacement part. On other occasions a client will request custom hydraulic fluid tanks, and pipe assembles made for specialty projects. In rare cases you will get cracked equipment that must be repaired. This niche is a good addition to any welding shop but most of the work is done on a per piece basis. For example, a local hydraulic hose manufacturer might give you fifty pieces to weld every month at $20.00 each. It is not a lot of money, but it can pay your rent. On the other hand, if you take this niche to the extreme you might just have a specialty business that does nothing but weld hydraulic fittings all day long, and ship them nationally and internationally.

Another option is you can manufacture some specialty items and sell them in catalogs or through local businesses. For example, instead of welding fittings for a local hydraulic hose manufacture, sell them directly to the customer. You also have the options of making your own product line if you have the capabilities to machine parts.

Finally, you can buy used hydraulic equipment and remanufacture it yourself. After that you can either sell it online or through dealers. There is a lot of money in remanufacturing specialty items because it costs the clients a lot less than purchasing shiny new parts.

Heavy Equipment Repair

Heavy equipment always needs repairs and service. There are three areas in this niche that you can service:

- Hard Facing
- Fabrication
- Buying and Reselling Heavy Equipment

Ask any heavy equipment business owner what their

first worst nightmare is, and they will tell you not having enough work, but second on the list is always 'repairs'. If you are able to offer hard facing repairs, you will be saving those business owners a ton of money! Buckets wear out, tracks get smooth and metals wear thin. The only other option is buying a new part that will wear thin just as quickly. Business owners know that hard facing adds a protective coating and surface area that lasts longer than the original manufacturer's base metals.

Fabrication of new and specialty heavy equipment parts are big business. To gain that type of business the clever welder will do it cheaper than their competitor. However, in most cases the competitor is the manufacturer and the prices are not exactly a bargain. The bottom line for most customers is the bottom line, period.

If you have some money and don't mind a small gamble, then you can buy used heavy equipment, parts and accessories that are broken or salvaged for pennies-on-the-dollar to repair and remanufacture for resale. You may work at your own pace and time, but you are risking your money on something that may take a while to sell. However, it is a very profitable business and it has made many people millionaires.

Industrial Construction and Maintenance

The two most obvious places for welding jobs are in

industrial construction and maintenance. Here is the problem; most companies that get this type of work have deep pockets or have personal connections. That leaves small welding businesses in two places. First you can subcontract or you can staff yourself as a welder. Neither one of these options is a great prospect, but if you need the work it is an option.

The down side of this business goes back to chapter 6 and networking. Most of the contracts in this area are signed amongst friends. That means golf buddies, and people the decision makers personally know. It does not matter who does a better job or offers the better price. It all comes down to relationships.

The only expectation you might allow for is if you happen to have a project manager friend, or any decision maker as a pal. In that case you can charge anything you want and do what you want. Welcome to the world of politics and relationships. It is closer to home than you expected.

Jungle Gyms, Slides, Swings and Kid's Parks

Here is an awesome business that many people would kill to get a piece of (it's too bad they don't weld). Jungle gyms and kid's parks are popping up all over the place and they are mostly made from metal pipe (steam pipe fittings) and have a lot of profit in them. There is a lot of work to be had in this area. There are two possible paths:

- Custom/Made-to-Order, On-Site Fabrication
- Pre-Made, Ready-to-Install Equipment

Custom, on-site equipment is the better paying of the two types of work in this field. Being an expert in this niche can land you some pretty large contracts and the best part is that there is not a lot of competition. You can get these types of contracts from:

- Advertising Your Services
- Sub Contracting
- Bidding On Federal, State, County, City and Township Jobs

Expect to travel quite a bit if you go the route of building custom made jungle gym equipment. Many of the people who do this type of work not only travel out of state, but sometime overseas. Some buyers of these types of services are:

- General Contractors
- Sub Contractors
- Cities, States, Counties, Towns and Parks Departments
- Resorts and Hotels
- Schools
- Cruise Ships
- Churches
- Shopping malls

- Amusement Parks, Carnivals and Circuses

Pre-made jungle gym equipment can be a good source of income for fabrication shops. It gives you a product that you can sell directly and control what type of work you are willing to do. There is some competition in this area but anyone who is able to create a cheaper or better product (kid's will be the judge of that) will be successful.

Some ideas of where you can find buyers are:

- Sell To Hardware Stores
- Online and In Print Catalogs
- Offer a Reseller Program
- Sell Directly Through Advertising, Commercials or a Website
- Contractors and Builders
- Cities, States, Counties, Towns and Parks Departments
- Schools
- Churches

You will likely find a following if you fabricate quality jungle gyms and their components. If you do it right and contact the right purchasing department, you may just land buyers such as Lowes, Home Depot and True Value Hardware. If so, then you will get more work than you will know what to do with! Just be careful not to over extend your business to the point of failure.

Kitchen Accessories and Custom Fabrication Services

Kitchens are big business in the retail world and there is a demand for custom accessories that are high end. If you have the market to build high end fixtures for kitchens then you have a niche business. How many welding businesses do you think are servicing high end kitchens? I'm guessing there can't be too many of them out there.

This is an excellent niche market full of many buyers that have ideas, concepts and dreams that no one is wanting to service. Look at the Food Network channel and all of the cooking magazines at the supermarket store shelf. How many people with money do you think that you can do something for? Your buyers can come from:

- Architects
- Home Builders
- Contractors
- Magazines
- Celebrity Chefs
- QVC or The Home Shopping Channel
- Kitchen Cabinet Shops
- Kitchen Accessory Stores
- Super Markets

Motorcycle and Chopper Building, Modification and Custom Parts Production

You need to get into the custom motorcycle business if you like playing with motorcycles, want to build choppers or create custom accessories. There is no shortage of custom bike builders but there is still enough work if you are willing to do it for the right price. It is even possible to go 'global' if you have a very creative design that no one can resist because of it's uniqueness. The catch with building and modifying bikes is you need to be creative.

There are many opportunities out there but can you make the customer happy? Bike design is an art form that is not yet tapped and customers are craving change and more affordable prices. Think about this; Could you spend 60k on a custom bike? The answer is 'probably not', but you would be a buyer if it were only 25k, right? This is a supply and demand issue and most Americans want a custom bike. Give the market what it wants and you will be rewarded.

Nuclear Power Plant Welding Services and Staffing

Nuclear power plants are always in need of welders. The issue is that you need to pass some really strict welding certifications because the visual inspection is critical. Additionally, there will be considerable safety training and a requirement for a red badge along with a clean background.

Aside from all that, these types of jobs are very easy to

do and often you will find yourself getting paid for very little actual work. You would basically be staffing out your employees and yourself, or hiring people for the job if you do decide to go this route. Most welders on these job sites rarely strike an arc because everyone is always sitting around waiting on paper work to clear. Safety and procedure are at the highest importance at these jobs sites.

The money is good, the competition is small and the work requires a lot of travel. All it takes is a few calls to contractors that do this type of work and in time you will have a thriving business.

Off-Road Vehicle Modification and Repair

Like playing with toys? Off road vehicle repairs and modifications are a good business if you live near and service the right area. The down side is most people who will want to hire you don't have a lot of money, but will be grateful for your work. You can get into this field through the following channels:

- Advertise as a Specialty Service
- Go to Race Tracks
- Meet ATV and Off Road Vehicle Dealer
- Mail Flyers to Local Racing Teams
- Create a Unique Product
- Follow or Join Local Off Road Enthusiasts Clubs

The fabrication and welding of off road vehicle modifications is best done for people who are racing or have a business that allows them to resell your work. You can go two ways in this niche. You can either do custom work or create aftermarket products. Some examples of aftermarket products that you can create are:

- Handle Bars
- Racks
- Grab Rails
- Suspension Kits
- High Performance Exhausts
- Exhausts That Are Quiet (for a change) or Silencers
- Loading Ramps
- Custom Trailers

You have a great chance at selling your product if you can improve a vehicle or add a function that customers would want. In this community most of the businesses are small to mid-sized so there is always a need and want for more specialized products.

Oil Industry and Natural Gas Welding Services

Everybody knows that energy is big business and welding is one of the services most needed in the energy segment. There is a ton of work in the oil and natural gas industries specifically. As a small welding business you won't be getting a big piece of the pie. Exxon, Mobil,

British Petroleum and all of the big names have that market. You have no idea about the oil industry, it's history or how it works if you don't know about Standard Oil and John D. Rockefeller. The reality of working with the energy industries is that you have a chance to win small scraps of work from the big boys, but not much more.

However, bear in mind that what these industries consider scraps can, in fact, be **big business** for a small welding or fabrication business. First of all, the big oil companies don't build their own equipment or run pipe lines. They subcontract the work out to oil service contractors. What you need to do is find the companies that are awarded the contracts to build or maintain their equipment. The oil industry hires welding businesses in three ways:

- Sub Contracting
- Rig Welding
- Staffing

These are the only options you get to pick from in order to work with the energy industries. There is a lot of money to be made for people who have the skills to fill these jobs. You should know that you will be expected to:

- Be real good at what you do
- Pass a background check
- Pass a physical exam
- Pass a drug and/or alcohol screening

- Leave cell phones in the car or at home daily
- Be willing to travel extensively and regularly

Subcontracting is done by identifying the companies that get the oil industry contracts and then waiting for the jobs to be posted and bidding on them. The jobs can range from staffing welders as your own to building equipment. Additionally, being a mobile rig welder using your own truck to weld on pipe lines is a viable option. This industry is regulated by the American Petroleum Institute or A.P.I., and the code under which you must certify is theirs. They typically weld downhill and run an E6010 most of the time.

The building of equipment is a different story in that they need to have perfect welds. This industry is closely regulated and codes are strict. But you would be amazed at how low the standards are for what pipe lines are allowed as far as root penetration. This is a fact that you should research very well.

Rig welding is very common in the oil and gas industry. Its workers are best described as independent contractors rather than full time employees. Companies that need welders and equipment look for rig welders. Rig welders are welders that have their own truck and welder that they weld with.

A rig welder gets paid by the hour and gets two rates; the rate for welding and the rate for their equipment. For example, you get $35 an hour for welding and $15 an hour for your machine. That brings the rate to $50 an hour. Be warned though that rig welding can be a double edged sword.

You can make an awesome living if you have your own equipment and it also happens to be a good market. On the other hand, if things are slow you'll soon be in trouble if you cannot afford to make the heavy duty loan payments on your nifty equipment. For example, at one point the rig welders of Oklahoma were earning $150 an hour so every welder went out and invested $70,000 on a welding rig. When the oil prices fell, rig welders were suddenly working for $10 an hour. Barely enough for gas money. That means owning a large debt, with no possibility of paying it back. Be sure to investigate all the variable before committing so such a large sum.

Finally, you can always staff oil field jobs with yourself or your employees. Staffing is a safe business and it is a way to get a lot more money out of a company than actually working for them yourself There are a lot of staffing companies that send out really bad welders, but all the contractors want is someone who can do the work correctly. Just give them what they want and you will have more work than you can handle.

Piece Work Shop, Welding and Repairs

A welding piece work shop is considered hitting a home run to anyone who owns a welding shop. You get welding work in the door and once you are done you send it right back. It is a simple and clean business that allows you to weld and nothing more. Most of the work is TIG welding and for the majority of welders, that is the most satisfying type of work. This type of welding business is perfect for a one person operation and requires minimal welding equipment and shop space. Some examples of what types of work you can do are:

- Machine Parts Welding
- Aircraft Part Repairs and Welding
- Marine Part Repairs and Welding
- Automotive Parts Repairs and Welding
- Hydraulic Parts Welding
- Tool and Die Repairs and Welding

There are endless possibilities for a small welding shop. The work can come from one industry or many depending where you advertise or network. For example, you can go around to small machine shops and get jobs that pay by the piece. You pick up the pieces and when you are done welding them, you deliver them back and collect a check. Another example is specializing in repairing cracked automotive cylinder heads, engine blocks and manifolds (requires specialized knowledge of alloys and pre-heating/cool down procedure). If your area supports small executive airports, then you can refurbish aircraft parts for established repair shops or even start buying and reselling them yourself.

If you go about it the right way, you will have a specialty

that people in your state, even internationally, will need and they will even gladly ship work directly to you to make sure it gets done efficiently and right. Another possibility is that you have a diverse range of local customers giving you many types of jobs.

The one thing you need to remember about getting this type of work is to find businesses that are small and do not have a welder on staff. Once a business has enough work to hire a full time welder, there is no longer a need for them to continue paying you 4 to 5 times the hourly rate.

Quality Control, Weld Inspection and Training

Working as a welding inspector is **_easy_** compared to welding all day long. Welding and fabrication businesses of all types and sizes will need inspectors eventually. The work comes either from doing weld inspection, writing weld procedures, certifying welders or even training or hiring welders.

You should be aware that you will need to have a minimum amount of experience (depending on the codes your working with) and you also need to pass the CWI or Certified Welding Inspector exam, which most people fail unfortunately. The CWI exam is very expensive at approximately $5,000.00 or more per test, pass or fail. However, if you pass the test, you can market your inspection services to **_any_** business that deals with welding or fabricating metal products.

After passing the CWI test you can offer on-site welding certification tests to businesses and even schools (some welding instructors are not CWI's) for $99 a test (pass or fail) with a five test minimum. You could also hunt down all of the local industrial projects and offer your weld inspection services, as well as consult for them on their welder hiring practices. As a welding inspection business, there are a lot of options.

Some weld inspection businesses specialize in radiograph testing (x-ray testing) while others only work on weld procedures (paper work). The choices are yours as long as you know how to reach potential customers. Always do your research first and get to know your market and potential client base before jumping in.

You are only one small exam away from being a CWE or Certified Welding Educator if you have a CWI license. You should know that the federal government offers grants (free money, not loans) to people who want to open a welding school. Welding schools are extremely profitable if you market the school properly. Many schools are failing because they never mastered this area. Think about this; 20 students a year, each paying a tuition of $20,000 comes out to $400,000 a year in gross income. There is more than enough money to support a welding school as long as there is no shortage of students. There will ALWAYS be new welding students.

Railroad Equipment Fabrication, Repair and Modification

Railroads and their customers frequently need a great deal of metal fabrication. Everything is made of steel and many companies need custom made shipping jigs and container modification. You can offer your services directly to the railroads by bidding on their jobs through their purchasing department or you can advertise your services to customers that have special shipping needs.

You will be amazed at all of the specialty shipping equipment that is needed and there is no place to get it made. For example, if you had to ship a large odd shaped object, how would you pack it? The answer is building a set of braces to support it inside the container. That is the only option. Another profitable area is found in restoring old locomotives and cars. You may need to travel around the country or world but there are businesses that need these services.

Restaurant Equipment Manufacturing, Fabrication and Repair

Want to be involved in a business in which there is always high demand *and* pays well? Then the restaurant equipment industries are a good choice. It does not matter where you live, there are always restaurants and they need lots of stainless steel work done. The work ranges from custom fabrication of shelves to cooking hoods and general stainless steel repairs. The price tags for these types of equipment are very high and a lot of the work is made-to-order. The one main driving factor of this industry is the turnover of new restaurant owners. Earlier in this book I wrote about buying equipment for your business, and the predators that profit from new business owners. Here you have a chance to be that predator or simply service that industry. Some places to get work are:

- Restaurants Owners
- Restaurant Supply Stores
- Restaurant Equipment Stores
- Advertising in Restaurant Supply News Papers
- Advertising in Business Broker Publications
- Buying and Reselling Used Restaurant Equipment

There are a lot of opportunities in this field, all you must do is market your services.

Shipbuilding, Repairs, Modification and Welding Services

The ship industry provides many opportunities for you to provide your welding services. There are tons of heavy plate and pipe, lots of wear and tear on the vessels, and the owners have deep pockets that don't care what it costs as long as it is done right and right now. The shipping industry can be broken into two groups:

- Shipbuilding/New Construction
- Ship Repairs and Service

Shipbuilders have lots of work and a global economy just keeps feeding them more as we keep trading with other countries. Ship builders hire welding businesses in three areas:

- Sub-Contracting
- Welding Services
- Staffing

Subcontracting is done by bidding on the jobs that the primary ship builder has to offer. It's just like subcontracting industrial projects. You contact the purchasing department and bid on the jobs. The work can range from simple jobs like railings that the shipyard does not have the time or personnel to build all the way up to modular ship components ready to install.

Depending on the shipyard and their customer's requirements, they are in many cases, required to subcontract a percentage of their work to small businesses. If they don't comply by the terms of their contract on this point, they may lose the next contract.

Shipyards have a business that has many ups and downs. This makes it difficult for them to estimate how many employees they need all year round. If a shipyard gets a lot of work then they are in desperate need of welding and staffing services. Depending on the shipyard you can be hired as a contractor with your own equipment to weld in their yard with your own employees (just like a rig welder). This also opens the door for staffing your employees for their needs at whatever hourly rate you want to charge. Most shipyards staff and hire contractors all of the time. It just does not make sense for them to hire too many full time employees because once the job is done they must pay unemployment benefits to those employees.

Ship repair and servicing is better suited to smaller welding businesses. Most jobs in the repair area are done by small businesses or independent contractors. Jobs can range from welding cracks to replacing hull material. Many of the jobs in the repair and servicing industry are done while the ship is sailing out at sea. For a shipping company or ship owners, the down time is just too expensive to dock a ship at a port for a few weeks to get the job done. The only case of a ship staying at port is if the repairs or modifications require cranes or there are additions being made.

For example, ships are dry docked every so many years to repaint the hulls and while at dry dock all major modifications and repairs are completed. It is just like a industrial shutdown where time is very critical. This opens up opportunities for independent contractors and small welding businesses to get some really well paying contracts. The way this type of work is found is by marketing and networking your business to local dock masters or shipping companies. This is a word of mouth type of business and you need to work your way in by meeting people.

Some of the more common types of work in this field include pipe replacement, ballast tank rebuilding, hull and deck reinforcement, ship additions and major ship modifications. A common practice in this industry is for multiple small contractors to join together to do one big job. This is something that is very rare in other industries and it just tells you how small this community is.

Stainless Steel Welding and Fabrication

Stainless steel is a niche business that is considered both a specialized field and a very general welding business. This niche, just like welding aluminum, is a way to convince your potential customers that you are the expert in this field. Simply put, you get the trust of a niche business but the marketability of a general welding shop.

Welding stainless is really nothing special except for spreading heat, bracing joints, and that the cutting off of material is done by plasma cutters, but inevitably customers like to feel as though they succeeded in hiring a stainless steel pro. This also allows you to charge more per hour and earns you more trust than your competitors.

Tool and Die Repair, Modification and Welding

Tool and die repair is a very profitable business for welding shops. There is a lot at stake and the customers are always praying that you can get the job done. The reason that they need you so badly is that you are saving them a lot of money in repairs and down time.

Tool and die repair is very much like fixing cracked engine blocks, heads and manifolds. There are special alloys used, preheating and stress relief techniques are a must, cleaning and joint prep is not easy, and the customer has a lot riding on getting the job done properly. If you fail then they are going to lose a lot of time and money on getting a new die made.

For example, a single die can easily cost in excess of $100,000.00 and if you charge $5,000.00 for the repair that takes a day to complete, then you will have not only saved the customer a lot of money, but the shortened down time could be worth millions of dollars to their business.

This type of work is found by advertising and networking your business to the right industries. You can specialize in one or advertise to multiple industries. For instance, you can find work from any manufacturing business, machine shops, or manufacturing equipment dealers. The industry that needs you does not know it until something breaks and they have a large order to fill. It is just like fixing cracked engine blocks an hour before a big race. The customer does not know that they need you until they find out the cost and down time of getting a new engine block.

Under Water Welding

Underwater welding is not exactly a niche welding business because it falls under commercial diving, but let's assume that you are a welder who also happens to have a commercial diving license. You can easily make some serious money welding under water. There is always a demand for underwater welders and inspectors. It is a high risk job that relatively few people around the world are capable and qualified to do.

Obviously, the type of work that will be available to you will depend on where you live and how far you are willing to travel. Oil companies are always in need of quality underwater welds, bridges need to be inspected and nuclear power plants need inspectors to make sure everything is working fine. This is and is not a welding niche business, but you will be well compensated if you fall into this category.

Vessel Fabrication and Repair

Vessels are high priced items and require a certain amount of skill to build or fix. This is a strictly American Society of Mechanical Engineering (ASME) code business that requires large performance bonds and insurance. The up side is that a small, skilled shop can land some really fat contracts from engineering and procurement businesses.

Pressure vessels are used in many industries and many welders are not able to weld them up to code. Besides that, almost everything is x-rayed which puts a lot of welding shops out of the game specifically on the root of a weld. If you are capable of building pressure vessels properly, all you need to find is the right customers.

Waste Container Repair and Fabrication

Garbage and recycling businesses are a great place to get work. They always have equipment breaking down, containers that need welding and trucks that need steel replacements. You can even build better equipment and sell to them directly if you have a shop. But the work does not stop there.

Many of these garbage disposal and recycling businesses are building power plants that burn garbage and turn it into steam power for producing electricity. Once you get into this field you can expand a lot more than you think because of the green energy trends that have been established. Some ideas of services to offer are:

- Container Repair
- Hard Facing
- Custom Equipment Fabrication

- Steam Pipe Installation
- Truck Body Repair and Modification

Subcontracting and servicing are the two ways to break into the waste industry. Subcontracting is basically biding on jobs posted by the purchasing department, while servicing is done by marketing and networking your business with the right people. That is about all there is to getting involved in the garbage business.

Niche Business Summary

Simply put, you will need to create or identify some sort of niche that requires servicing if you want to own a successful welding business. Even general welding shops sell their services to niche industries. The only question is, "What niche(s) are you going to service?"

This portion of the book has been an exercise in learning how to be a successful business owner. Welding business owners who are successful often don't know how to weld. They are experts in selling their services to other business owners. The reality about owning any business is that marketing is necessary in order to make deals. If you can't do that then you are an employee.

This is exactly why most welders can't understand why management does what it does. Management is only interested in profit and nothing else. As a business owner your thoughts will be directed to getting the job, then making money, and finally to paying your welders. The business-to-business transaction is your only goal, and that is the thought process that it takes to survive, like it or not, fair or not. These are the facts; employees make 20% of the money for 80% of the work because 80% of the work comes from 20% of the initial sales effort.

In the end you need to decide which is a better fit for you. Are you better off simply welding for someone else? Or are you ready to commit yourself and your resources to owning and managing your own welding business? Are you prepared to reap the corresponding rewards for the risk and effort that it takes to be a successful welding business owner?

Hopefully this chapter on niches was an eye-opener for you and you learned how and where to promote your business. Business owners make their money by making themselves valuable to other business owners who need their services. The goal is to find people who are willing to pay a lot of money for what you have to offer and don't know where else to turn for their needs.

CHAPTER 11 - BUSINESS PLANS AND WHY YOU NEED ONE

If you Fail to plan, then you must plan to Fail

Writing a business plan to run your proposed business is the equivalent of fabricating from a blue print or drawing. You have an end product with a clear vision of what you want to build, the materials needed, a general idea of the cost and time required, and the steps it's going to take to complete it. Most business that fail have no plan or idea of what they want to do. Most welders who open up a business expect to buy equipment, put out a sign and then the work will magically come to them.

What typically happens is the so-called-business-owner ends up buying a bunch of personal toys (welding and fabrication equipment) and then goes back to working for someone else because they are broke or can't get any work. How many welders have you known who've have a welding business but have wound up working at some job to make ends meat? Everyone knows a few of these types of "business owners". Heck, you may already be personally guilty of this all-to-common practice.

A business plan is a blueprint that defines what your business is about. It is a mental exercise in managing your business. How will your services satisfy the needs of your customer? Answer that question! It defines your business and works as your work schedule by setting goals and time lines to reach to your goals.

Finally, it helps you determine the budget for how much you will need to spend to achieve your goals. Name your primary customer, and what will enable you to have a long term relationship with that customer. Name the niche that you have chosen. What marketing tools will you employ? How much exactly will you spend on them? Your plan must address all aspects of your business and its future.

In the end, a business plan is where you set your goals, know your budget, and finally get your business off of the ground. It is amazing what a business plan on paper will reveal about your business before you spend a dime. The more time you spend on your business plan, the greater the chances of your business being successful. It will give you a clear direction to follow by *analyzing* your planned actions *before* you put any money or time behind your ideas. What are you going to produce? How should we do that, and what equipment do we need?

Who Will Be Looking at Your Business Plan

The first person that needs to look at your business plan is you, and you need to keep reviewing it just to stay on course. But know that there are other people who will see your business plan. Some people who should review or need to see it are:

- Partners
- Operations Managers
- Potential Customers
- Banks and Lenders
- Investors

There is more to writing a good business plan than to simply create a blueprint for yourself. It can be used to give a partner or an operations manager your vision of and goals for the company, as well as what is expected of them.

If your business is involved in manufacturing or dealing with publicly traded companies, your potential clients and/or investors are going to require your business plan and financial statements. The same applies when you are bidding on jobs or selling to national companies. They want to know what your plans are for the future of your business and how it may affect them.

Usually, your clients will be concerned that you will be in business long enough to deliver what you agree upon should you be awarded their contract.. In many cases, they will also want to see your financial statements. It is just a matter of a business protecting itself and avoiding unscrupulous businesses.

You will NOT get past the loan application process without a solid business plan, if you're going to need to borrow money for equipment, basic start-up costs, or operation capital. No matter what you tell them, they want to see it on paper. You can kiss that financing goodbye if your business plan looks weak, sloppy or not well thought through because no bank will lend unless they know you plan to invest the money wisely.

Your business plan will always be the crucial selling point to any investor looking to give you the money your business needs. Investors are interested in ideas that can almost guarantee the investment won't be lost. Most investors will judge your business by the business plan itself. Let's say you have a product that you want to bring to the market. In most cases you will be able to locate many investors and fill out their required forms. The one thing they will inevitably require from you is your business plan. <u>Typically, investors will not meet with you or even take any calls unless they have already seen your business plan.</u>

Business Plan Basics

A business plan should contain eight main sections of information. The eight sections are:

1. Summary of Your Business and its Goals
2. Company Description of What Your Business Does
3. Market Analysis of Your Services/Products and Competition
4. Management and What Type of Business Structure You Use?

5. Services and Products
6. Marketing, Networking and Sales Plans
7. Start-Up Money Needed or Financing Requirements
8. Loan Request and Pay Back Plan (if needed).

Most business plans want you to look forward 3 to 5 years. This is an excellent time line because most businesses take about 5 years to get properly established. After that point the chances of your business failing are small.

The summary of your business and its goals is just that, a summary. It is a brief description of what your business does, who it is run by, management's experience, products and services offered, your financial situation and where you want your business to be in 3 to 5 years.

The company description and what your business does is nothing more than the niches you plan to service. The person reading it should know who your potential customers are, what exactly you are offering, why someone should hire your business instead of a competitor's and what is unique about your company.

The market analysis of your services, products and competition needs to be detailed. This comes back to why a customer should choose you over the competition that has been in business longer than you. Some of the information you need to include is how big the market is that you are servicing, what percentage of the work you think you can steal from the competition, how stable the industry is, how profitable your business is going to be, what incentives you intend to offer, and what are the potential obstacles to growing your business.

The management and business structure of you business is more about who you are and what you have to offer. The business structure is pretty simple and all you need to do is fill in the blanks such as a corporation, limited liability company, or sole proprietorship. Management is a list of all of the people who are management, what their rolls are, experience, what their salary or pay is and what they bring to the company.

The 'Service and Products' section of your business plan is a detailed list of what your business has to offer. Are you a mobile welding service or do you manufacture a product line? This is where you define what your customers are going to pay you money for. Perhaps you plan to specialize in tool and die welding. Maybe you own a patent and have the exclusive right to manufacture a new type of a expansion joint. It could be that the shop will only be dealing with remanufacturing aircraft parts.

You should also include any future products or services that you are working on and how much work your business can handle. For example, if Wal-Mart approached you to help build all their California stores, you need to know in advance that you can actually handle that work load. Guessing won't cut it and could cost you a lot in the way of money and reputation. You could also be developing a new type of interlocking handrail system that many hardware stores are eager to stock. This is where you have a chance to dangle that carrot and "WOW" your investors and clients.

Marketing, networking and selling your products or services is the core of any business and you need a solid plan to follow through on. This portion of your business plan is not only needed for other people to look at, but it is also where you set your planned schedule and specific goals. The bottom line is all about what are you going to do, specifically, to get the business in the door?

How you're planning to market, network and sell your products is what any bank, investor, partner, potential client and anyone who has an interest in your business needs to know. It also happens to be the one part of your business plan that you need to follow more faithfully than any other! You can slack on other areas from time to time, but as long as you have and *follow* a solid marketing plan you will succeed. If you don't stick to the plan, then you really should polish up your resume because you will be working for someone else very quickly. It's *that* important.

If you are going to need a loan or get an investor involved, they are going to want to know precisely how you plan to use their money to create a return on their investment. You need to make a detailed list and description of why you plan to use their money as well as how and when you plan to pay them back.

I do want to point out that there are many services from which you can purchase a business plan template or you can have one made for your business. There is also another option that is much better. It is the Small Business Administration or the SBA. They will help mentor you through the process of organizing and writing your business plan all the way through the point at which you feel your business is successful. They will literally sit down with you in person and help you with whatever you need and give you a lot of very valuable advice. The advice will come from a successful business owner that already know your nich. There is no cost to you. This is all paid by the government to help grow our economy. You can visit their site at **www.sba.gov** and you can find any type of information that you are looking for on their site for free. It can be worth tens of thousands of dollars in free advice and counseling.

Below is a short list of places to get help from or to buy a premade business plan.

www.SBA.Gov

www.BPlans.com

www.BusinessPlans.com

www.LawDepot.com

www.MasterPlans.com

www.WiseBusinessPlans.com

CHAPTER 12 - HIRING HELP AND EXPANDING YOUR BUSINESS

Options for Hiring Temporary Help

Any business you are able to name depends upon it's partners and employees to make it successful. You really need to weigh your options if you need to hire help. Going about it the wrong way can land you in some serious legal troubles and it may even cost you your business. Remember that the better the quality of people you hire, the better the quality of your business.

Before you advertise, or post a job, you must create a job description. You should explain the duties, the responsibilities, and the requirements/skills that you require. Here are some ways you can hire help:

- Partner the Job or Contract
- Sub Contract the Work
- Borrow a Employee from Another Welding Business
- Use a Staffing Company
- Hire an Independent Contractor
- Internships
- Full and Part Time Employees

Partnering a job or contract is the easiest way to deal with a short term overflow of work. It does not require any serious financial or liability commitments on your part and it keeps the costs down. Besides that, you may just be saving someone's business, home, car, truck or equipment from getting repossessed. This goes back to the chapter about buying or renting equipment.

There is always another business that is struggling and if you give them a part of the contract then you can save a lot of money and you also get someone grateful for the work. Besides, if the job is too big for you, they may have the equipment and people that you need for no extra cost. It is a win-win situation.

Think about sub contracting the jobs out if you get a lot of work and can't handle the load. Most welding businesses have no clue how to get work and will be more than happy to get some work to help pay the bills. Again, this is an area that carries no real legal liabilities or long term financial commitment on your part. Once the job is done you are free and clear.

Additionally, you can make a killing offering other welding businesses work for a smaller amount than your signed contract if you don't feel like working and can get the jobs in the door. In the real estate and business broker world this would be considered flipping or assigning contracts. It is a common practice.

Let's say you have a fabrication or manufacturing shop and all of a sudden you have a huge inflow of orders. Instead of hiring employees, you can make an employee exchange with a local business owner that is simply trying to keep their employees working. You reap the benefits of not having long term liabilities, and in an instant, you have trained staff ready to go. The business owner will be more than happy to keep their people working and at the same time, earn a small fee.

For example, if you find a business that has three employees that are inactive, but the owner is still paying them. You then have them work in your shop, for a small mark-up in their hourly rate. This puts food on the table of the workers, keeps the other business owner in business, and saves the other business from paying unemployment benefits. In the end, you get an employee without any of the legal and financial issues that would otherwise come with an employee. Vice versa, you can offer your workers to another shop to keep them busy if you have a slow streak, and employees to pay.

Staffing companies are both a good and bad choice for filling jobs. First, they will fill the job but their workers may not possess the skill level that you require. The upside is you don't have any real legal liabilities on your part. The down side is they will send just about anyone they can in the hope of staffing the job. It is a gamble but there are some really good staffing companies that can fill the job without any commitment on your part and it can be done at a very reasonable rate.

Independent contractors are the hybrid between employee and contractor. They work like employees and earn a little more per hour, but get the benefits of a contractor. That means no insurance, drug tests, time and a half, double time or unemployment benefits. On a personal note, I feel that independent contractors are the future of construction jobs. It gives the worker personal responsibility and the company what they want without any legal issues. You can't beat that and it is very common in Europe and the Cruise Ship industry.

Internships are the best source of cheap employees. If you can offer real world experience and training for students, then you may just have some worthwhile free labor on your hands. This is what Hollywood does and many industries want to accomplish. They want a desperate worker that has nothing to lose and needs experience in the field. It is the hard working employee that has a lot to prove and needs your approval on a resume.

Full and Part Time Employees

To start, if you need to hire employees you need to know what kind to hire. It seems there are two types of employees these days. The first is the type that works for small businesses and the second is those who work for major corporations.

Small business employees are hard workers that usually do what they are told. They never question management or dare to sit around waiting for instructions. As I remember, working for small businesses there was a saying I'd hear occasionally, "if you are not welding or fabricating, then you are sweeping or cleaning." You made sure you were doing something at all times!

Welders who work for big businesses never do anything more than welding. If you are a welder then, that is all you do. As an employee of a big business I was shocked to weld only 20 minutes out of a 12 hour day. The welders knew their place and everyone knew what they can get away with. The difference between a small shop and a major corporation was the quality of the work.

Big corporation employees worked a lot less but had superior skills. Small business employees worked hard but did not produce the quality wanted or needed. As a business owner, I would never hire a big business employee because they are no different than union workers. They feel skill and commitment justifies their lack of corporate commitment.

As a small business you want to hire employees who have a past working for other small businesses. If not, you will quickly find that big business employees are not willing to work hard and just want to ride the clock. Not good for you. Check backgrounds and stick to employees who are accustomed to working in small companies. A good employee will do what management says, with no questions or requests. So now ask yourself which type of employee you really want to hire.

Full and part time employees are a big financial and legal liability. This is especially true when it comes to welding, fabrication and manufacturing industries. To begin, you need to do the following checks on any employee:

- Skills Tests
- Drug and Alcohol Tests
- Criminal Background Check
- Prior Employment Verification
- Workman's Compensation Claims
- Optional: Psychological Screening
- Optional: Credit Check
- Optional: Medical History
- Optional: Driving Record
- Optional: Vision and Hearing Tests
- Optional: Physical Exam

The list is pretty long but it is about you protecting a business from a bad employee that can literally take a company down. This is an area that can get a business in a lot of financial and legal trouble if you hire the wrong employee.

For example, there are a lot of employees that are only interested in working just long enough to get fired thereby getting an unemployment check out of you. Others have serious mental issues and could have the potential to physically hurt other employees or customers. Victims can and will come after you for financial compensation.

Remember, you are legally on the hook for anything

that an employee does on your company time. Prepare to face major consequences should one of your employees causes an accident while under the influence of drugs or alcohol. You should know that with these types of employees, it really is only a matter of time before they drag you and your company down with them, The bottom line is people who need work are going to lie and do whatever it takes to convince you to hire them. Whatever happens after that is your problem.

I do want to make a personal note here because I have been through all of these checks (yes, I passed every test and even the strictest). I feel many of these checks are a direct violation of the United States Constitution because they are an unreasonable search into your personal life and it is not right. But in the end you need to think like a business owner and your employees are nothing more than an investment in your business. You need to treat employees in such a way that they know their place. Otherwise, you will lose control and they will run your business into the ground. We all know how that ends.

Skills tests are a given because you cannot trust an applicant's description of welding skills. This is a hands-on business and if you can't pass a simple welding, fitting or machine set-up tests then you don't deserve the job! Why risk hiring someone who can't pass a simple test like a vertical up fillet weld?

Drug and alcohol testing is a big issue these days. In most cases they are required by the insurance company that will be issuing your workman's comp policy. Besides that most employees that have drug or alcohol issues are a serious liability to the business.

Criminal background checks are required by most government and state contracts. Besides that, how far are you really willing to trust an employee that comes with criminal background? The best advice is get a full criminal background search done, know who you're thinking about bringing on board, and avoid even larger entanglements later on.

Checking references was the old way of hiring employees but these days companies are so big that they have no clue of who was or was not an employee. As mentioned earlier you should avoid employees that have a history working for big businesses. They want to much money, will not work as hard and in most cases will refuse doing other duties like cleaning toilets. No matter what you decide, make sure you check their background and what is on the application matches what you researched.

Workman's compensation insurance is a huge cost to the business owner because of the employees who take advantage of the system are BIG trouble. You and your insurance company need to know immediately if an employee is taking advantage of your workman's compensation insurance. It is a major liability that can literally take down your business.

Psychological screening is an optional choice, but crazy

people don't warn you. Most military, nuclear and Government jobs require a psychological test.

A Credit Score is an excellent gauge by which to determine a person's trustworthiness. Many businesses will not deal with other people or businesses that have bad credit. Credit affects almost every type of insurance, lease and just about any area that is needed to get an idea about how responsible they are.

Checking someone's medical history is a good way to avoid anyone who is trying to get you on the hook for workman's compensation. You need to remember there are many people out there looking for a business to sue or pay them to stay home. These employees see you as the enemy and in most cases just don't want to work. Besides that, providing health insurance to a scam artist of this nature will also lead to a huge increase in health insurance costs for the entire company.

Driving records and DUI checks are a must for anyone who plans to give an employee their vehicle to drive. First you need to know that the person is licensed to operate the type of equipment you need operated. Then you need to know if this person is responsible. For example, three DUIs and one hit-and-run over the last 25 years should have you running away from hiring them. Finally, you need know whether your insurance company will insure them to drive for your business.

Vision and hearing tests are sometimes needed. For

most welders a vision test should always be given. Many welding certifications require them. In some cases the welders' weld might not be as good simply because they are not aware that their vision has worsened. Hearing tests would only be needed if the employee is going to be involved in work that would require good hearing. Both vision and hearing tests are a good idea and can also improve someone's life if they did not know they needed help.

Physical exams are becoming a standard part of hiring. This is for insurance reasons, company protection and ensuring the employee can handle the job. A physical should always be required of employees that are working in extreme conditions, far from any medical help or in situations that require good physical health in general. For example, many defense contractors require it, and if you plan to work offshore, you will not go anywhere until a doctor examines you. As a new business, the last thing you need is a employee claiming their bad health was caused in any way by your business.

Getting Financing, Loans, Investment Deals or Bringing on Partners

Before beginning you need to know that business loans and financing options change on a daily basis and this chapter is just to give you some basics. You can get some of the best and up-to-date information, advice and help from the Small Business Administration (SBA). There are literally hundreds of books about financing, investing and business related loans. Some people have Doctorate Degrees on this subject, so just view this chapter as a basic guideline.

You have a few options to choose from if you need money for your business. Unfortunately, all of them will cost your business in either profit or control. This is an area that you need to be very careful with as far as who you choose to get involved with, and most importantly what you agree to and sign for. Almost everyone you talk to in this area is going to be looking to take advantage of you. Everyone in these types of businesses is either wanting for you to sign your life away or they see that you have something of value (including your family's assets) and are looking for a way to get it from you. Your options for getting capital are:

- Business Financing or Loans
- Bringing on a Partner or Investors

Business Financing or Loans

Business financing or loans are made by banks, commercial lenders, brokers and private lenders. As a borrower looking for a business loan there is a lot that you will need to know. To begin with and in order to save you time *read the next sentence carefully*. Most lenders do not deal with new businesses unless you have something of value that will secure the loan; collateral.

An example of collateral is, you may have a house and car that are paid off and you are willing to hand them over as collateral against the money you are borrowing. If you you don't pay the loan back, they keep your collateral as payment. Or perhaps you have a contract with a buyer and the amount you need to finance is just the money necessary to make the product itself so that you can deliver it and get the payment for you services. This would be considered a bridge load. The bottom line is that no bank or lender is going to risk lending any money unless they have something of real value to pay for a loss if you don't make it. Unlike personal or home loans, most business lenders are going to ask for the following information:

- Business Plan
- Your Personal Background (Qualifications, Education and a Resume)
- Collateral or Assets to Secure a Loan (House, Vehicle or Anything of Value)
- Bank and Financial Statements (Personal and Business)
- Credit Report (Personal and Business)
- All Legal Documents for the Business

As you can see, most lenders want to know everything about you, your business and anyone related to you. They are only willing to lend to businesses that have a high probability of loan repayment. Chances are, if you get a loan or some sort of financing, it is because they had you sign away everything you own and there is very little risk in the loan. Since you are now a business there is no consumer protection. You need to be smart because there are predators looking to take advantage of you. Use a lawyer or a skilled accountant. That is the nature of business loans.

Having a Partner or Investor

Having an investor invest in your business is a great alternative. Most investors who believe in what you are doing will give you all of the money you need, their time, advice and expertise to make sure you succeed. Investors come in many forms but the most common are:

- Private Investors
- Investment Banks, Hedge Funds and Portfolio Managers
- Shareholders

Private individuals can be found by either advertising or word of mouth. They can be anybody who has money to invest in your business including family and friends. The deal requires that you sign away a percentage or portion of your business in return for using their money to get your business what it needs.

Another commonly used term for individual investors are hard money lenders. These are common in the investment real estate market. The down side of getting a private investor is that they may want to start running your company, or in many cases, are only looking for an easy way to own your business and get rid of you.

One of the issues of being a new business owner is knowing whom to trust, and you never know if a investor is genuinely trying to help you or is trying to cash in by steering you down the wrong path. Most individual investors that will lend money are sharks and are looking for a quick kill that is already bleeding. This is business, so be very careful about what you agree to, and what you sign. Always have the papers checked by your attorney *prior* to formally executing them.

Investment banks, hedge funds, and portfolio managers are always looking to find great start-up businesses to invest in. If you can get their attention, you can feel confident that THEY think your business has a high probability of making it big. All these people do is invest other investor's money for them into businesses, products and ideas that they feel are going to change the world and make a lot of money.

Keep in mind that these investors really know what they are doing and only go after the best deals they know about. For most welding related businesses you would need to have a welding-related product or service that could potentially change the world in order to get backing from these investors. On the flip side, you will have gained a great deal of support should you succeed in getting their financial backing.

Think about it this way, almost every corporation that you know to be successful today was funded by one of these well-funded investors. GE, Microsoft, Ford, Facebook, Google and thousands of other companies are among such companies. These investors are the best in the world and they know what works.

Finally, you may want to take your company public if think you just might have the "next big thing" up your sleeve. What does that mean? That means selling shares or stock (same thing) to investors on a stock exchange. This means your company gets a ticker symbol like GE has "ge" or Fluor Corporation (formally Floral Daniels and before that it was known as Daniels) has "flr".

The way this is done is by contacting a registered stock broker/investment banking firm, or underwriter, and telling them you want to take your company public. If they like your company (a good story that sells is all they care about) they will talk. The upside of going public is that you can raise millions of dollars and you never need to pay it back.

The stock price of your company is all that investors look at. On some exchanges there are very minimal legal requirements and you don't even need to disclose anything close to what banks require. If your business does well, the stock holders will see their stock prices go up because other investors want to buy shares. If not then your shares will become worthless. The best thing about having a publicly traded company is that whenever you need money you just sell some shares. It is like printing your own money.

The most important to remember about going public is that if you happen to fail you still get to keep you home, car and whatever your family owns plus the money the broker got for your business. The way you can find investment bankers to take your business public is by searching for "pink sheet stocks underwriters/brokers" or "taking your company public".

These brokers can take your business to the exchanges that allow and deal with unregulated and undisclosed companies to sell shares. It's a good option but you could lose control of your company, or even get bought out by a bigger business if you have something of value.

Another positive feature of going public is that your business has national exposure, which typically means bigger and more profitable contracts. If you are wondering about big exchanges like the New York Stock Exchange (NYSE) or National Association of Security Dealers Automated Quotations (NASDAQ), all you need to understand is that those exchanges only deal with well established businesses that have lots of money in the bank. Chances are that is not you.

Buying and Merging with Other Businesses

If you are just starting out and can afford to buy a business that is successful then do it before starting one from scratch. Buying an established business is a sure fire way to make sure you do not lose your shirt. As a business owner you need to find opportunities from other businesses. Some examples are:

- Buying Used Equipment and Acquiring Employees
- Buying Entire Businesses and Customer Bases
- Merging Struggling Businesses with yours for the Lion's Share of Profits or Running Them into Bankruptcy and Then Taking Over Their Assets

As a business owner the troubles of other unsuccessful businesses make great opportunities for you to improve your business based the miscalculations and misfortunes of other business owners. This is a time where you can buy equipment for pennies on the dollar. Your competitor's entire business, patents, product rights and customers can be yours for next to nothing.

Finally, you can buy a struggling business and either run them into bankruptcy by any for an easy kill to get more customers, or you can merge with them and then drive the owner out because you can get them to sign an agreement that favors your views because they have nothing to negotiate with.

These are the realities of running a business as a real life business owner. The question is, what side of the deal do you want to be on?

Buying Patents and Trademark Production Rights

To start, you need to call a lawyer if you have a product, trade name or right to produce a product. You need to protect your property from other shops selling your work. Owning certain rights puts competitors out of business and ensures your company stays strong. On the other hand, you can buy the rights to produce products for your own company if you have some money to invest.

Saving money to buy the product rights or innovative product is an excellent plan if you cannot create one products yourself. Patents, trademarks and product rights are all you need to own a successful welding, fabrication or manufacturing business. Financing and many other business opportunities will come to you if you own these rights to the correct products.

As a matter of fact, you don't even need to actually manufacture the product if you just get the right patents, trademarks, etc. You can simply sit back and collect royalties. For example, you might own the patent to a specialty door hinge. Anyone who wants to sell that door hinge will need to pay you a percentage of the price basically forever so long as your door hinge is sold.

No one can make a product that you have designed and named if you have a trademark without your express permission. Finally, if you own any of the above, there is no competition and you can charge whatever you like for your product. As a new business owner you may struggle to find the right idea, but there are many people out there that have a good idea but don't know what to do with it. If you, as a business owner can do both, and can close a deal, then you have it made.

Product rights can be found through business brokers. There are always people retiring from a business and are looking to sell the rights and customer bases that they have spent a lifetime building. As a buyer you get someone's life work and next to guaranteed income from their work. If you do things right, you can build a manufacturing empire that is foolproof by searching for the right products to own. All it takes is cash and finding the right person at the right time. You can find these types of opportunities by searching for "business brokers, patent's for sale, business for sale" and so on. It just takes searching in the right places to find them.

Business, Patent and Product Brokers

Here are some useful places to find information and search for products.

www.BizBuySell.com

www.IdeaConnection.com

www.BizQuest.com

CHAPTER 12 - RESOURCES TO HELP WITH YOUR BUSINESS

Government Help, Free Advice, Mentors and Minority Owned Businesses

As a new business owner you have many places to turn to for advice. There are many agencies and government programs that will not only give free advice, but also get you financing and even a mentor. The government knows its survival rests in the creation of small businesses. The risks are huge, but the payoff is enormous for the government.

The primary government agency involved in aiding and assisting the small business owner is called the Small Business Administration or SBA. If there is anything you need and can't afford in advice, the SBA has you covered. They have lots of programs for all types of businesses that will help you. The SBA has assistance for the following and a lot more:

- Business Plans
- Mentoring
- Partner and Board Members
- Financing, Loans and Investors
- Government Grants
- Minority Business Assistance

- Legal Help
- And Just About Anything Your Business Will Need

The SBA is in almost every city and town in the United States. Even if you live in a rural area, chances are your closest town has an office. The SBA is truly interested in helping you have a successful business because their success is based on growing our economy.

You will hire workers if you succeed, and if you hire workers, you will pay taxes that pay for the SBA. Without the SBA this country would fail because small businesses would not stand a chance.

www.SBA.gov

Summary and what you should have learned from this book.

Reading this book should have been an exercise in thinking like a *business owner*. As a welder your thought process is nowhere near the same as a savvy business owner. We think of the quality of the weld and fabricating a perfect product. Everyone one of us has ***never*** understood why management does what they do, until we understand the business world. Hopefully, this book has opened your eyes to what it takes to run a successful welding business (or any business), and at the very least point you in the right direction. In welding, fabrication and manufacturing we learn by practicing with a trial and error method. There is always a person we can ask for help in the worst case scenario, but business and being a business owner is different.

Owning a business and competing for the available work out there is like being in a war-like environment. No one is out to help, everyone is trying to under bid and undercut you, let alone finding someone offering you help of any kind. Over my years as a welder I have met many people who at one point started their own welding business, but were soon working for someone else. I am guilty of this myself because I was in the same situation and those jobs paid the bills.

What changes a welder into a business owner are their priorities and what they hold valuable. Welders value the quality of the work, but entrepreneurial business owners value the amount of the pay check, and the psychic reward from doing it all themselves. Just remember, that businesses that do bad work sometimes get many of the jobs. Most customers have no clue about what a good weld is and it is all about salesmanship and marketing.

As a business owner it all comes down to getting the work in the door and cutting costs as much as possible. It is survival and the exact opposite of being a welder. A welder gets paid for welding and working hard. A business owner gets paid for working the welder hard and closing the deal. As a business owner you get 80 plus % of the profits for less than 20% of the work.

The bottom line is if you want to own any type of welding related business, you need to know how to be a sales person. Your job is marketing and meeting people who might need you. If you focus on those duties then your business will fly. If you can't sell then you need to buy product rights or anything that allows you to compete. If not, you will fail and become an employee again.

As a business owner you need to get the work and find ways to beat your competitors. It is financial warfare and your side must win! Find the way to take the work from your competitor, to cut costs and all expenses and work your employees harder than the other person is doing. Only by doing this will you maximize your return.

You must remember that without a customer, you are just someone who knows how to weld. Understand your customers' needs, be well acquainted with your customers, do whatever you have to do to keep your good customers. Handle any complaints or problems in a very prompt and decisive manner.

Do not let your daily business tasks make you lose your laser- like focus on one thing, the customer. The customer is everything to your business. You are nothing without a customer. You must understand their needs, have a relationship with and keep the customer, and promptly take care of customer problems. Remember who pays you and your employees, who creates demand for your services, who decides when your company expands? This is the reality of the business world. It is just the way the world works and it is what it takes to survive as a business owner. Business is business; if you want a friend get a dog....

ABOUT THE AUTHOR

The Welding Business Owner's Handbook was written by David Zielinski who is the creator and owner of www.GoWelding.Org. David is a journeyman pipe welder, fabricator and a successful business owner.

Special Thanks to:

Stephen Maurice Jacoby for editing this book. Stephen is a former Vice President of Chemical bank and a expert in the corporate fitness field. Steven holds a M.S. from N.Y.U. and a M.A. from The New School. He is retired but finds retirement to be less than challenging and is always looking for corporate projects to give him some sanity from the everyday boring grind.

Finally, Miss Ballou who has been my best friend and confidant for more than a decade, and contributed in more ways than one to this book. Miss Ballou is a public school Social Studies and Language Arts teacher who was formerly a corporate head hunter, marketing assistant, human resources manager and franchise operations manager. She is also the person who originally got me writing. Visit her site **www.MissBallou.com** if you are interested in reading for fun or want to get your kids, or a kid you know hooked on reading. But be warned! Kids who have no interest in reading and take her advice about great books to pick up, won't be able to put her book suggestions down.

Index

5080393R00130

Made in the USA
San Bernardino, CA
22 October 2013